Springer Theses

Recognizing Outstanding Ph.D. Research

Aims and Scope

The series "Springer Theses" brings together a selection of the very best Ph.D. theses from around the world and across the physical sciences. Nominated and endorsed by two recognized specialists, each published volume has been selected for its scientific excellence and the high impact of its contents for the pertinent field of research. For greater accessibility to non-specialists, the published versions include an extended introduction, as well as a foreword by the student's supervisor explaining the special relevance of the work for the field. As a whole, the series will provide a valuable resource both for newcomers to the research fields described, and for other scientists seeking detailed background information on special questions. Finally, it provides an accredited documentation of the valuable contributions made by today's younger generation of scientists.

Theses are accepted into the series by invited nomination only and must fulfill all of the following criteria

- They must be written in good English.
- The topic should fall within the confines of Chemistry, Physics, Earth Sciences, Engineering and related interdisciplinary fields such as Materials, Nanoscience, Chemical Engineering, Complex Systems and Biophysics.
- The work reported in the thesis must represent a significant scientific advance.
- If the thesis includes previously published material, permission to reproduce this must be gained from the respective copyright holder.
- They must have been examined and passed during the 12 months prior to nomination.
- Each thesis should include a foreword by the supervisor outlining the significance of its content.
- The theses should have a clearly defined structure including an introduction accessible to scientists not expert in that particular field.

More information about this series at http://www.springer.com/series/8790

Jack Dunger

Event Classification in Liquid Scintillator Using PMT Hit Patterns

for Neutrinoless Double Beta Decay Searches

Doctoral Thesis accepted by
the University of Oxford, Oxford, UK

Author
Jack Dunger
Merantix AG
Berlin, Germany

Supervisor
Prof. Steven D. Biller
Department of Physics
University of Oxford
Oxford, UK

ISSN 2190-5053   ISSN 2190-5061   (electronic)
Springer Theses
ISBN 978-3-030-31618-1   ISBN 978-3-030-31616-7   (eBook)
https://doi.org/10.1007/978-3-030-31616-7

This Springer imprint is published by the registered company Springer Nature Switzerland AG
The registered company address is: Gewerbestrasse 11, 6330 Cham, Switzerland

For Jane and David

Supervisor's Foreword

The search for neutrinoless double beta decay is one of the highest priority areas in particle physics today; it could provide insights into the nature of neutrino masses (currently not explained by the Standard Model) as well as how the universe survived its early stages. One promising experimental approach involves the use of large volumes of isotope-loaded liquid scintillator, but new techniques for background identification and suppression must be developed to be able to reach the required sensitivity levels and clearly distinguish the signal. Results from this thesis constitute a significant advance in this area, laying the groundwork for several highly effective and novel approaches based on a detailed evaluation of state-of-the-art detector characteristics. The thesis is extremely well written and includes a particularly clear and comprehensive description of the theoretical motivations. The effective use of diverse statistical techniques is also impressive and well described. The signal extraction framework developed is very professionally constructed and contains clever algorithmic solutions to efficient error propagation in multi-dimensional space. This code is being used as the foundation for future signal extraction work within the SNO+ collaboration. In general, the techniques developed in this work will have a notable impact on the field.

Oxford, UK
October 2019

Steven D. Biller

Abstract

The SNO+ experiment is the successor to the SNO neutrino detector, which replaces its heavy water target with a liquid scintillator one. The primary physics goal is the search for neutrinoless double beta decay ($0\nu\beta\beta$) in ^{130}Te, which will be loaded into the scintillator. Fitted with >9300 photo-multiplier tubes, the SNO+ detector will have the highest photo-cathode coverage of any large liquid scintillator detector. This thesis shows that, at this light collection level, SNO+ is sensitive to differences in the scintillation pulses produced by electrons, positrons and gammas, and that these differences may be used to classify single-site $0\nu\beta\beta$ events and multi-site radioactive backgrounds which emit γ. This pulse shape discrimination technique (PSD) is applied to background events from radiation originating outside the detector, which limit the experiment's fiducial volume, and potential internal radioactive decays, like ^{60}Co, which are otherwise difficult to distinguish from $0\nu\beta\beta$. A new signal extraction framework is described and used to perform 2D fits in energy and event radius, which estimate an expected limit on the $0\nu\beta\beta$ half-life of $T^{0\nu}_{1/2} > 1.76 \times 10^{26}$yr, at 90% confidence, assuming an exposure of 4.0 tonne·yr of ^{130}Te. The corresponding limit on the effective Majorana mass is $m_{\beta\beta} < 49.7$meV, using the IBM-2 nuclear model. Further, it is shown that adding PSD as an additional fit dimension can reduce the SNO+ 3σ discovery level on $m_{\beta\beta}$ from 190meV to 91 meV, assuming the same exposure. The final portion of this work discusses what more could be achieved using a liquid scintillator experiment which can separate scintillation and Cherenkov signals in time. A simulation of a SNO+ style detector, filled with a slow scintillator and equipped with a high coverage of fast, high quantum efficiency PMTs is used to demonstrate separation of Cherenkov and scintillation signals and reconstruction algorithms for electron and $0\nu\beta\beta$ events are described. Differences in Cherenkov signals are used to distinguish $0\nu\beta\beta$ from the solar neutrino elastic scattering background, and to demonstrate for the first time that, in principle, the $0\nu\beta\beta$ mechanism may be determined in liquid scintillator by fitting the angular separation and energy split of the two emitted electrons.

Acknowledgements

First, I'd like to thank my supervisor, Steve Biller, for his tuition over the years and for giving me the space and confidence to try things out. I'm indebted to Jeff Tseng for teaching me to write software, which was at least half of my work on SNO+, and Armin Reichold who showed me how to work in a lab and the optimal technique for all-you-can-eat sushi.

I had the pleasure of sharing the Oxford SNO+ student office with Luca Cavilli, Krish Majumdar, Jeff Lidgard, Esther Turner, Tereza Kroupova and the unquenchable Chris Jones. Thank you for the coffees, the chats and the proof reads —and for letting me quiz you all about your work, which was somehow always more interesting than my own. Extra thanks to Ed Leming for beers and sage advice during the final throes of the Ph.D.

The work presented in this thesis builds on the hard work of others in the SNO+ collaboration and their tuition, insight and professionalism contributed to all aspects of it. I feel lucky to have worked with so many people I admire. Among many others, Matt Mottram, Morgan Askins, Ian Coulter, Rob Stainforth, John Walker, Billy Liggins and Rebecca Lane made the long cold days in Sudbury something I would look forward to.

I'd also like to thank everyone who works in the Denys Wilkinson Building for making it a great place to come to work. In particular, Kim Proudfoot and Sue Geddes, administrators in the department, got me though in one piece, solving everything SNO+ could throw at them over the years[1]. Farrukh Azfar has been a dear friend, for his incredible roast lamb and tireless curiosity about the big questions that got me into physics.

Last but absolutely not least, Louisa thank you for keeping the world turning around me these past months, all the while doing a more difficult job across two cities. You've been wonderful.

---

[1]After one mix up, we needed to re-book a flight to Sudbury, via Copenhagen, 2 hours before I was due to leave!

Contents

Chapter 1
Introduction and Theoretical Background

1.1 Neutrinos

The neutrino was first proposed to save energy conservation in nuclear β decay. By 1914, Chadwick had shown that the β energy spectrum was continuous [1], but β decay was then thought to be a two body process, in which the energy of the β and recoiling nucleus are exactly constrained by energy conservation. What followed was many years of controversy, as people questioned the apparently impossible experimental results [1]. They were finally confirmed by Ellis and Wooster in 1927 [2] and Meitner proved that the missing energy could not be accounted for by neutral γ rays [2]. These results famously led N. Bohr to suggest that perhaps energy conservation applied only in a 'statistical sense' [3].

In 1930, in a letter beginning 'Dear Radioactive Ladies and Gentlemen', Pauli proposed that perhaps a neutral, weakly interacting particle was created alongside the β, carrying off the remaining energy without detection [4]. He concluded that the mass of the particle should be 'the same order as the electron mass, and in any event not larger than 0.01 proton masses'. The neutrino would have spin 1/2 in order to resolve the apparent non-conservation of angular momentum in these decays. He lamented that he had proposed a particle that could not be detected, something 'no theorist should ever do'.

In 1934, E. Fermi developed his theory of weak interactions [5, 6], modelled by a single 4-fermion vertex that included the neutrino. Its subsequent success in describing β spectra convinced most that neutrinos existed, but detection was thought impossible after Bethe and Pierls estimated the cross-section for $\bar{\nu} + p \to e^+ + n$ to be $<10^{-44}\,\mathrm{cm}^2$ at $1\,\mathrm{MeV}$ [2].

That was until, in 1956, Cowan and Reines achieved the first detection of $\bar{\nu}_e$ using positrons created in inverse beta decay interactions on 1400 litres of liquid scintillator [7]. Using a nuclear reactor as an intense $\bar{\nu}$ source, they were able to overcome the tiny neutrino cross-section to produce a measurable event rate. For this work, Reines was awarded the 1995 Nobel prize in physics.

© Springer Nature Switzerland AG 2019
J. Dunger, *Event Classification in Liquid Scintillator Using PMT Hit Patterns*,
Springer Theses, https://doi.org/10.1007/978-3-030-31616-7_1

Discovery of the other two neutrino species came later. The interactions of the muon neutrino were detected in 1962 by Lederman, Schwartz and Steinberger [8]. The tau neutrino was only proposed after the discovery of the τ lepton and not observed until 2000 by the DONUT collaboration [9].

Meanwhile, the Fermi theory of weak interactions had failed to explain several phenomena, including β decays where the nuclear spin changed by 1 unit, provoking considerable debate on the exact algebraic form of the weak interaction [2, 10]; it would require an experimental breakthrough to make progress.

In 1956, Lee and Yang began to question what most had considered a self-evident fact: that the laws of nature should be invariant under parity transformation [10]. That same year, Garwin, Lederman and Weinrich [11] demonstrated that parity was violated in μ decay and Wu [12] showed it was violated in the β decay of ^{60}Co. Wu and her collaborators cooled ^{60}Co atoms to 0.001 K and used a magnetic field to align the nuclear spins. They then measured the orientation of the β produced in ^{60}Co $\rightarrow$ ^{60}Ni decay and found that electrons were preferentially emitted anti-parallel to the parent spin. The spin is invariant under parity transformations, but the electron momentum is not, so an asymmetric distribution proved parity violation. Spin conservation also implied that the weak interaction preferentially produced left handed electrons, a fact that led Lee and Yang to argue for a two component model of the neutrino, in which neutrinos/anti-neutrinos are massless and always left/right handed [10].

After Wu's experiment, support for this model came from the V–A theory of the weak interaction, developed by Feynman and Gell-Mann [13], Shudarshan and Marshak [14] and Sakurai [15]. It predicted that the correct form of the weak interaction produced maximal parity violation, a property that was naturally expressed in the lepton sector using the two component neutrino model.

Corroborating experimental evidence came from the Goldhaber experiment [16]. It measured the helicity of neutrinos produced in electron capture on ^{152}Eu:

$$e^- + {}^{152}\text{Eu} \rightarrow {}^{152}\text{Sn}^* + \nu_e \tag{1.1}$$

By measuring the polarization of γ emitted by the relaxing Sn nucleus, they were able to establish that the ν_e spin was always aligned opposite to its momentum, regardless of its direction. In effect, they had found that the ν helicity was the same in all inertial frames, to within the error of their experiment. If this was always true, the neutrino had to be massless.

As a result, the two component model of the neutrino was codified into the standard model of particle physics (SM), which went on to explain all experimental data on strong and electro-weak interactions, until the discovery of the neutrino mass. To date, the neutrino mass is the only terrestrial observation of physics beyond the standard model (BSM). The rest of this chapter is devoted to this discovery and its implications.

1.2 The Neutrino Has Mass

The discovery that neutrinos have mass essentially came from two experimental conundrums: an observed deficit in ν_e from the sun and, later, an observed deficit in atmospheric ν_μ. The theory of neutrino flavour mixing could explain both phenomena, but required that the neutrino had finite mass. The theory was experimentally confirmed by the ground-breaking results of the SNO, Super-Kamiokande and Kam-LAND collaborations, leading to the 2015 Nobel prize for physics.

This section outlines the evidence for neutrino flavour oscillations, setting out what is known and, as yet, unknown about the neutrino masses and the parameters that determine oscillation probability.

1.2.1 The Solar Neutrino Problem

The standard solar model (SSM) is the modern theory of the sun's energy generation. Its development began between the late 1920s [17] and culminated with the work of Bethe [18] and others [19], who proposed that the sun's intense energy generation is powered by the fusion of hydrogen into helium, proceeding via a chain of intermediate reactions in two cycles: the pp and pep chain (Fig. 1.1) and the CNO chain. By the early 1960s, there was a consensus that thermonuclear fusion fuelled the sun and that the pp chain was the dominant contributor [20], but there had been no experimental proof.

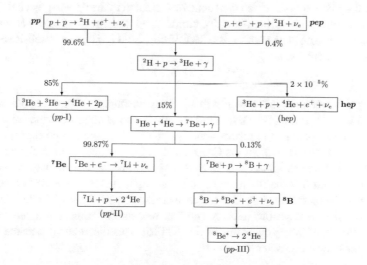

Fig. 1.1 The pp and pep solar fusion cycles [21]. The percentages are branching ratios. Five of the processes depicted emit neutrinos, named hep, pp, pep, ^{7}Be and ^{8}B (bold face)

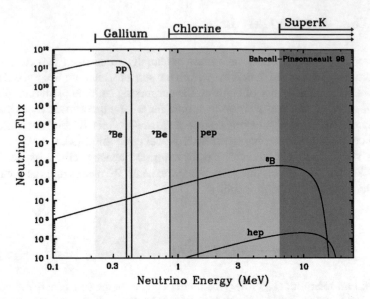

Fig. 1.2 Solar neutrino fluxes at the surface of the earth [27]. The flux units are cm^{-2} s^{-1}. The annotations above the plot show the range neutrino of energies that produce events above detector threshold for a range of technologies

In 1964, Ray Davis and John Bachall proposed that the model could be verified by detection of ν, produced in the pp chain [22, 23]. Present day expected fluxes of each of these neutrino types are shown in Fig. 1.2. The experiment they devised was first suggested by Pontecorvo; it detected neutrinos created in the pp and pep chain, using neutrino capture on chlorine (Eq. 1.2), stored deep underground in the Homestake mine, South Dakota. The reaction's 0.8 MeV threshold [24] made it sensitive to ^{8}B, pep and CNO neutrinos.

$$\nu_e + {}^{37}\text{Cl} \rightarrow e^- + {}^{37}\text{Ar} \qquad (1.2)$$

They published their initial results in 1968. By chemically separating out the ^{37}Ar produced in the reaction, they measured a capture rate of solar neutrinos of 3 SNU[1] [25, 26]. The impact of the results was significant: they confirmed thermonuclear fusion in the sun and ruled out the CNO cycle as the dominant mode, for which the contemporary flux estimation was 35 SNU [24, 26]. Furthermore, they inferred the central temperature of the sun to be 16×10^6 K, again matching the SSM prediction [24]. For these successes, Ray Davis was awarded the 2002 Nobel Prize. However, it also provoked one serious tension: the flux measurement did not agree with the SSM prediction of the pp cycle flux of 6 SNU [26]. This discrepancy became known as the solar neutrino problem.

---

[1] 1 SNU (solar neutrino unit) is equal to the flux that produces 10^{-36} captures per second per target atom.

Years later, the discrepancy gained even more attention when large water Cherenkov detectors were built to search for proton decay. These detectors had thresholds of >5 MeV so they also detected ^{8}B neutrinos. The Kamiokande experiment measured the ^{8}B neutrino flux over 8 years to be around 50% of the predicted SSM flux, a 2σ discrepancy [28]. Their result agreed with the Homestake measurement using completely independent detector technology. Later confirmation came from the Super-Kamiokande experiment. Their first results published a ^{8}B flux of less than half of the SSM prediction, again at 2σ [29]. The discrepancy only grew with more data; by 1998, Super-Kamiokande had measured the ratio of the ^{8}B flux to the SSM prediction as $0.358^{+0.009}_{-0.008}$ (stat) $^{+0.014}_{-0.010}$(syst), a result entirely inconsistent with the SSM.

Further hints were offered by gallium radio-chemical experiments. These used similar detection technique to the original Homestake experiment, with the key exception that the neutrino capture threshold for ^{37}Ga is 0.233 MeV [24]. At this threshold, the experiments were also sensitive to pp neutrinos, which were expected to comprise of over 99% of the neutrino flux. Moreover, the theoretical flux uncertainty for pp neutrinos was around 1%, rather than 10% for ^{8}B neutrinos [30]. The GALLEX/GNO [31] and SAGE [32] experiments each measured pp fluxes of $\approx$50% of the SSM prediction, both 5σ discrepancies.

Figure 1.3 summarises the results from the chlorine, water Cherenkov and gallium experiments [30] using the final rather than initial results from each. At the time, there

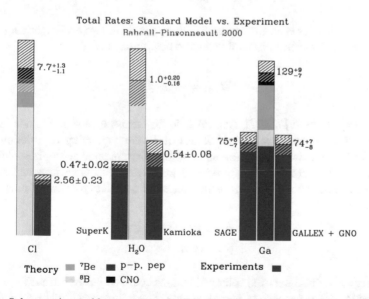

Fig. 1.3 Solar neutrino problem summary plot [33]. The left-most bar for the Cl comparison and the central bars for the H_2O and Ga comparisons show the solar standard model predictions without neutrino oscillations, the other bars show experimental measurements, the shaded regions indicate errors. For the central figure, the fluxes are given as a fraction of the standard solar model prediction. The left and right figures are in solar neutrino units

were three major inconsistencies. First, the left most plot shows that final ^{8}B flux measured at Homestake was around 1/3 of the SSM prediction. Second, although both the chlorine and water Cherenkov experiments measured a deficit, they were inconsistent with each other: if the ^{8}B-only flux measured in Super-Kamiokande was extrapolated to a capture rate at Homestake, the prediction *exceeded* the total ^{8}B + pep + CNO capture rate observed there. Finally, the event rate observed by the gallium experiments could be completely accounted for using only the predicted pp + pep fluxes and the ^{8}B flux measured by the water Cherenkov detectors. This left no room for the considerable ^{7}Be flux predicted by the SSM.

1.2.2   The Atmospheric ν Anomaly

Atmospheric neutrinos are created when high energy cosmic rays of energy 10^{8-20} eV [34] hit the earth's atmosphere and produce hadronic showers. Particles in these showers decay, ultimately producing a large number of the lightest charged hadrons, $\pi^{\pm}$. These decay to $\mu^{\pm}$ with 99.99% branching fraction and the emission of $\overset{(-)}{\nu}_{\mu}$ [35].

Below $\sim$1 GeV, almost all of these muons decay before they hit earth's surface producing $e^{\pm}$ and two neutrinos, one ν_e and one ν_{μ}.

$$\pi^{\pm} \rightarrow \mu^{\pm} + \overset{(-)}{\nu}_{\mu} \rightarrow e^{\pm} + 2\overset{(-)}{\nu}_{\mu} + \overset{(-)}{\nu}_e \tag{1.3}$$

At these energies, for each atmospheric ν_e there should be two accompanying ν_{μ} and the ratio of the detected ν_e to ν_{μ} flux should be *approximately*:

$$R_{\text{atm}} = \frac{\phi_{\mu}}{\phi_e} \sim 2 \tag{1.4}$$

However, the IBM [36, 37], Kamiokande [38] and Super-Kamiokande [39, 40] experiments measured much lower values of R, corresponding to a deficit of ν_{μ} events. These detectors used kilotons of water, instrumented with photomultiplier tubes to detect Cherenkov light produced in the detector. Neutrinos can produce high energy charged leptons via elastic-scattering and charged current interactions on water:

$$\nu_l \rightarrow l^- + W^{+*} \quad \text{CC} \tag{1.5}$$

$$\nu_e + e^- \rightarrow \nu_e + e^- \quad \text{ES} \tag{1.6}$$

The energy and direction of the charged leptons could be inferred from the quantity and direction of the produced Cherenkov light. These, in turn, could be related to the energy and direction of the incident neutrino [41].

Figure 1.4 shows these results, among others, in the form of a double ratio that accounts for energy dependence, neutrino cross sections etc. by comparing the data

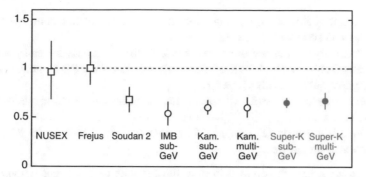

Fig. 1.4 Atmospheric ν_μ deficits from [42]. Squares represent iron calorimeter experiments, the circles are water Cherenkov detectors. The y-axis shows the flavour double ratio $(N_\mu/N_e)_{\text{data}}/(N_\mu/N_e)_{\text{MC}}$, where N_μ and N_e are the number of observed μ-like and e-like events in data and Monte Carlo, respectively

with Monte Carlo:

$$\frac{(N_\mu/N_e)_{\text{data}}}{(N_\mu/N_e)_{\text{MC}}} \tag{1.7}$$

Each of the water Cherenkov experiments and the iron calorimeter Soudan 2 measured a significant deficit.

1.2.3 Flavour Mixing

The solution to these inconsistencies came from the realisation that neutrinos change between flavours as they propagate. This occurs because neutrinos have a small but non-zero mass and the mass eigenstates they propagate in are misaligned with the flavour eigenstates that participate in weak interactions.

Neutrino oscillations were first considered by Pontecorvo in late 1950s in the context of active—sterile oscillations, by analogy with kaon oscillations [43]. The principle was later applied to the neutrino flavour states in 1962 by Maki, Nakagawa and Sakata [44], leading Pontecorvo to predict the solar neutrino problem in 1967 (before it was observed!) [45]. The standard oscillation probabilities used today were first calculated in the mid 1970s by Eliezer and Swift [46], Fritzsch and Minkowski [47], Blineky and Pontecorvo [2, 48]. Work on neutrino oscillations in matter followed in the 1970–1980s, principally developed by Wolfenstein [49], Mikheev and Smirnov [50, 51].

The theory of neutrino oscillations relies on the following arguments:

1. The neutrino states that couple in weak interactions $|\nu_{\alpha=\nu_{e,\mu,\tau}}\rangle$ are not the same as the mass eigenstates $|\nu_{k=1,2,3}\rangle$.

2. As the eigenstates of a linear operator, the weak states form a complete basis of the space describing the neutrino fields.
3. The weak eigenstates are therefore related to the mass eigenstates by a unitary operator U, representable as a square matrix U_{kj}, known as the Pontecorvo-Maki-Nakagawa-Sakata (PMNS) matrix.
4. If the masses are not perfectly degenerate, the mass state components will propagate differently.
5. So, even if a neutrino is produced in a state of definite flavour, its flavour will evolve as it propagates.

In essence, a neutrino created in pure flavour state $|\nu_\alpha\rangle$ will not remain in that state because its three mass components will accumulate phase at different rates. After time t, the neutrino state will have changed according to the time evolution operator $\hat{T}$:

$$|\nu(t)\rangle = \hat{T}(t)|\nu_\alpha\rangle \tag{1.8}$$

to give it a ν_β component of:

$$\langle\nu_\beta|\hat{T}(t)|\nu_\alpha\rangle \tag{1.9}$$

The probability of observing the neutrino as a different flavour, ν_β, after time t, is equal to the squared modulus of its ν_β component at that time:

$$P_{\alpha\to\beta} = |\langle\nu_\beta|\hat{T}(t)|\nu_\alpha\rangle|^2 \tag{1.10}$$

This probability is, in general, non-zero because $\hat{T}$ is not diagonal in the flavour basis. This must be the case because it is diagonal in the mass basis, and the two bases are misaligned according to U:

$$T_{\text{flavour}} = U T_{\text{mass}} U^\dagger \tag{1.11}$$

$$T_{\text{mass}} = \begin{pmatrix} e^{-iE_1 t} & 0 & 0 \\ 0 & e^{-iE_2 t} & 0 \\ 0 & 0 & e^{-iE_3 t} \end{pmatrix} \tag{1.12}$$

Here E_1, E_2, E_3 are the energies of the neutrinos three mass components, written in natural units.

In practice, many experiments are concerned with a beam of neutrinos created with definite flavour that propagate over some baseline, L. If oscillation between only two flavours is significant, as is often the case, it is sufficient to deal with a 2D mixing matrix:

$$U_{2D} = \begin{pmatrix} \cos\theta & \sin\theta \\ -\sin\theta & \cos\theta \end{pmatrix} \tag{1.13}$$

In this case, it can be shown that Eq. 1.10 leads to probability oscillations of the form [35, 52]:

$$P_{\alpha \to \beta} = \sin^2(2\theta) \sin^2 \left(1.27 \frac{\Delta m^2 L}{E}\right) \tag{1.14}$$

for Δm^2 in eV^2, L in km and E in GeV. So, provided the two neutrino mass states are not degenerate, and that they remain coherent, the probability of creating a ν_α and observing it later as a ν_β oscillates as a function of baseline L, with maxima at $L/E = (2n + 1)\frac{\pi}{1.27 \times 2\Delta m^2}$, $n = 0, 1, 2, 3....$ Critically, the $\sin^2 \theta$ term in Eq. 1.14 means that the position of the probability maximum depends only on the *magnitude* of Δm^2 rather than its sign.

The picture is complicated in matter by the Mikheyev-Smirnov-Wolfenstein (MSW) effect [53]. Elastic forward scattering interactions between propagating neutrinos and the surrounding electron cloud create an effective potential experienced by the neutrino. This alters the energy of its the mass eigenstates and the differences between them that determine oscillation probability. Importantly, these matter effects introduce probability corrections that are sensitive to the *sign* of Δm^2 as well as its magnitude.

If the neutrinos traverse a high but gradually decreasing electron density, as they would travelling through the sun, they can undergo *adiabatic conversion* above a certain resonance energy [53, 54]. This non-oscillatory phenomenon leaves initial ν_e in a pure ν_2 state for which the ν_e survival probability is constant:

$$P_{e \to e} = |\langle \nu_e | \nu_2 \rangle|^2 = \sin^2 \theta \tag{1.15}$$

Experimental results from Super-Kamiokande (SK), SNO and KamLAND confirmed flavour oscillations as the definitive explanation amongst many to the atmospheric neutrino anomaly and solar neutrino problem, using solar, atmospheric and reactor neutrinos, respectively.

First, in 1998, SK published a much more accurate measurement of the up-down asymmetry of the ν_μ flux [55]:

$$A_\mu^{\text{up down}} = \frac{N_\mu^{\text{up}} - N_\mu^{\text{down}}}{N_\mu^{\text{up}} + N_\mu^{\text{down}}} = -0.296 \pm 0.048(\text{stat.}) + 0.01(\text{sys.}) \tag{1.16}$$

where N_μ^{up} is the number of upward-going ν_μ events, coming from below the surface of the earth, and N_μ^{down} is the number of downward-going ν_μ, coming from the atmosphere above the detector. This 6σ deviation from 0 was conclusive proof that part of the ν_μ flux disappeared during paths through the centre of the earth. The corresponding ν_e asymmetry was consistent with 0 [55].

The explanation for this deficit is that atmospheric ν_μ oscillate into ν_τ as they free stream through the earth. Evidence for this came from a second SK result, measuring the flavour double ratio (Eq. 1.7) as a function of zenith angle.[2] Using this measurement, the SK collaboration were able to plot the ν_μ flux as a function of

---

[2]The smallest angle made with a vector normal to the surface of the earth.

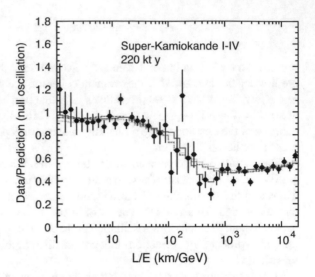

Fig. 1.5 Ratio of observed to predicted flux versus L/E for atmospheric ν_μ events at Super-Kamiokande [42]. L is the inferred baseline and E is the neutrino energy. Solid line: the best fit to neutrino oscillations. Dotted line: the best fit to neutrino decoherence. Dashed line: the best fit to neutrino decay

L/E for the neutrinos and compare with simulation. Figure 1.5 shows these results with a best fit to neutrino oscillations (solid line) and two competing explanations for the missing flux: neutrino decay [56] (dashed line) and neutrino decoherence [57] (dotted line). The sharp dip and upturn at around 500 km/GeV strongly favoured the oscillation hypothesis. The lack of anomalies in the ν_e data suggested the $\nu_\mu \rightarrow \nu_\tau$ oscillation channel, rather than the $\nu_\mu \rightarrow \nu_e$ channel.

The SNO experiment [58] was, like SK, a deep underground water Cherenkov detector. What made it unique was that its 1 kT target volume was filled with heavy water, D_2O. This change, proposed by H. Chen, allowed SNO to detect neutral current events via free neutrons produced in the break up of deuterium nuclei D:

$$\nu_x + D \rightarrow n + p + \nu_x \tag{1.17}$$

Free neutrons were detected via the γ emitted when the neutrons captured on H [59]. In the second phase of the experiment, the neutrons were captured on salt loaded into the D_2O [59][3] and in the final phase neutrons were detected with specially engineered neutral current detectors [60]. Crucially, this process is equally sensitive to all three flavours of neutrinos. This should be contrasted with the CC and ES processes, measured in SNO and light water detectors, which are only accessible to ν_e at solar energies. By measuring all three signals in the same detector, the SNO experiment was able to estimate both the ν_e flux and the total ν flux.

In 2002, the SNO collaboration published measured neutrino fluxes for ν_e and $\nu_{\mu,\tau}$ in its first phase [61]:

---

[3]Compared with D_2O, salt has a much larger neutron capture cross-section and produces a larger γ energy deposit. Both factors enhance the neutron detection efficiency.

$$\phi_e = 1.76^{+0.05}_{-0.05}(\text{stat.})^{+0.09}_{-0.09}(\text{syst.}) \times 10^6 \, \text{cm}^{-2}\text{s}^{-1} \tag{1.18}$$

$$\phi_{\mu,\tau} = 3.41^{+0.45}_{-0.45}(\text{stat.})^{+0.48}_{-0.45}(\text{syst.}) \times 10^6 \, \text{cm}^{-2}\text{s}^{-1} \tag{1.19}$$

The total neutrino flux was consistent with the SSM, and the ν_e flux was around 1/3 of the total flux, consistent with earlier experiments measuring a deficit. By providing evidence of non-zero $\phi_{\mu,\tau}$ at 5σ, SNO had categorically proven that neutrinos changed flavour and resolved the solar neutrino problem. Further, more accurate, results were published using data from the latter two phases. The final results shown in Fig. 1.6 were consistent with adiabatic conversion of the ^{8}B neutrinos inside the sun, with $\sin^2 \theta \approx 1/3$. The measured result was inconsistent with *exactly* 1/3, which ruled out very short baseline oscillations.[4]

However, there were still several possible alternative explanations for the *cause* of the flavour transitions, including non-standard interactions of massless neutrinos [62]. The KamLAND detector comprised of 1 kT of liquid scintillator, able to detect $\bar{\nu}_e$ via the same inverse beta decay interactions used by Cowan and Reines:

$$\bar{\nu}_e + p \rightarrow e^+ + n \tag{1.20}$$

When such an interaction occurred on nuclei in the target scintillator, the e^+ produced a prompt scintillation signal, followed 200 μs later by neutrino capture on hydrogen and the emission of a 2.2 MeV γ. This delayed coincidence was used to tag signal events to great effect.

KamLAND measured the disappearance of $\bar{\nu}_e$, produced by nuclear reactors, at a flux averaged baseline of 180 km [63, 64]. They were able to demonstrate vacuum oscillations by measuring the disappearance probability as a function of L/E (Fig. 1.7). The Δm^2 and $\sin^2(2\theta)$ measured in these vacuum oscillations was consistent with those measured by SNO, assuming adiabatic conversion. This was excellent evidence for neutrino flavour mixing.

The results of SNO and KamLAND also resolved the anomalies between the gallium, chlorine and water Cherenkov detectors. The inconsistent fluxes are now understood to result from the running of the ν_e survival probability with energy, and the resonance behaviour of the MSW effect in particular. pp, pep and ^{7}Be neutrinos sit below 2 MeV, where there is no resonant matter enhancement of the oscillations [2]. These neutrinos are produced over a region which is much larger than the oscillation length, so their survival probability is fixed at approximately the average for two-flavour vacuum oscillations, $1 - 1/2 \sin^2(2\theta) \approx 0.58$. On the other hand, some ^{8}B neutrinos have $E \gg 2$ MeV, so they undergo resonant conversion, with a survival probability of $\sin^2 \theta \approx 0.34$ [2]. For this reason, the average survival probability of detectable neutrinos differs between experiments with different thresholds, leading to apparent inconsistencies between their measured fluxes.

---

[4]If the oscillation length was shorter than the length of the detector, the oscillations would 'average out' to produce 1/3 of each flavour.

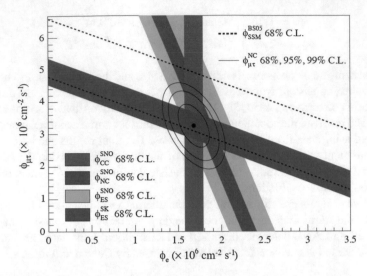

Fig. 1.6 Contraints on the electron neutrino flux, ϕ_e, and the combined muon/tau neutrino flux, $\phi_{\mu\tau}$, from the SNO and Super-Kamiokande experiments [61]

Fig. 1.7 Oscillations in the survival probability of reactor $\bar{\nu}_e$ as a function of L/E measured at KamLAND. L is the baseline from the source reactor to the detector and E is the anti-neutrino energy [64]

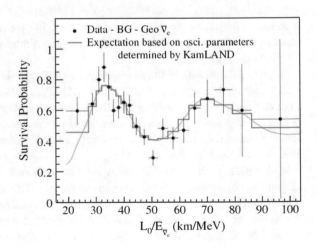

1.2.4 Oscillation Measurements

The flavour oscillation model set out in Sect. 1.2.3 contains many free parameters: the value of the components U_{kl} and the neutrino masses are not predicted by theory and must be constrained by experiment. By convention, experimental results are quoted in terms of the independent degrees of freedom of U and the mass squared differences Δm_{ij}^2 which are measurable in oscillations. The unitarity of the PMNS matrix means that it may be written in terms of three rotation angles θ and a single complex phase δ_{CP}:

$$U = \begin{pmatrix} 1 & 0 & 0 \\ 0 & c_{23} & s_{23} \\ 0 & -s_{23} & c_{23} \end{pmatrix} \begin{pmatrix} c_{13} & 0 & s_{13}e^{-i\delta_{CP}} \\ 0 & 1 & 0 \\ -s_{13}e^{i\delta_{CP}} & 0 & c_{13} \end{pmatrix} \begin{pmatrix} c_{12} & s_{12} & 0 \\ -s_{12} & c_{12} & 0 \\ 0 & 0 & 1 \end{pmatrix} \quad (1.21)$$

where $c_{ij} = \cos(\theta_{ij})$, $s_{ij} = \sin(\theta_{ij})$. Section 1.6 discusses the more complicated picture for Majorana neutrinos.

By convention, Δm_{12}^2 and θ_{12} are the parameters that control solar oscillations and Δm_{23}^2, θ_{23} produce atmospheric oscillations. The third pair Δm_{13}^2, θ_{13}, are associated with reactor neutrino oscillations. By measuring the L/E and magnitude of oscillation extrema, oscillation experiments measure each of the (Δm^2, θ) pairs and evidence from many neutrino oscillation experiments is combined into global fits of the neutrino data. A recent 2017 study by de Salas et al. [65] combines results from solar, atmospheric and reactor neutrino experiments as well as measurements from terrestrial neutrino accelerator experiments; each input is summarised in Table 1.1. Their fit has $|\Delta m_{31}^2|$ and Δm_{12}^2 constrained to within 1–2% and the mixing angles are each determined to 2–3%.[5]

There are still major unknowns in this picture. The first is the sign of Δm_{23}^2. Existing measurements of Δm_{23}^2 are dominated by vacuum oscillations which can only be used to infer its magnitude. The unknown sign of Δm_{23}^2 leads to two possible 'mass hierarchies' for the neutrino, depicted in Fig. 1.8. The de Salas fit prefers the normal hierarchy over the inverted hierarchy, but only with $\Delta\chi^2 = 2.7$.

There is hope for resolution of this question in the next two decades [83]. The accelerator neutrino experiment DUNE [84, 85] will conclusively measure the mass hierarchy using matter effects induced over a baseline of 1300 km. The RENO-50 [86] and JUNO [87] reactor experiments will directly measure the hierarchy using the interference between m_{13}^2 and m_{32}^2 oscillations in reactor $\bar{\nu}_e$ over intermediate baselines [88–90]. In addition, atmospheric neutrino experiments like Hyper-Kamiokande [91], IceCube PINGU [92, 93] and KM3Net-ORCA [94] can measure the hierarchy by inferring matter effects from the path dependent disappearance of atmospheric ν_μ.

The second unknown is δ_{CP}, which gives rise to CP violation in flavour oscillations of the form:

$$P(\nu_\mu \to \nu_e) \neq P(\bar{\nu}_\mu \to \bar{\nu}_e) \quad (1.22)$$

asymmetries of this type can be measured directly at accelerator experiments which are able to produce both ν_μ and $\bar{\nu}_\mu$, like T2K and NOVA. Contemporary experiments are not currently sensitive to δ at the discovery level; in the de Salas fit none of the region $\delta_{CP} \in [0, 2\pi]$ is excluded at 3σ. However, the planned large scale long baseline experiments HyperK [96] and DUNE [84, 85] can conclusively measure CP violation in the neutrino sector for a wide range of values of δ_{CP}.

---

[5]Though the octant of θ_{23} is still poorly constrained.

Table 1.1 Summary of neutrino oscillation experiments which have contributed to the 3-flavour oscillation picture at the time of writing

Source	Experiments	Measurement	PMNS Parameters	Reference
Nuclear Reactors	RENO, Daya Bay, Double CHOOZ	ν_e appearance	$\sin^2\theta_{13}, \|\Delta m_{ee}^2\|$	[66–68]
	KamLAND		$\sin^2\theta_{12}, \|\Delta m_{12}^2\|$	[63, 64]
Terrestrial Accelerators	MINOS	ν_μ dissapearance	$\sin^2\theta_{12}, \|\Delta m_{12}^2\|$	[69]
	T2K, NOvA	ν_μ dissapearance, ν_e appearence	$\sin^2\theta_{13}, \|\Delta m_{13}^2\|$ $\sin^2\theta_{12}, \|\Delta m_{12}^2\|$	[70, 71]
Atmopheric Neutrinos	Super-Kamiokande, IceCube DeepCore, Antares	ν_μ dissapearence, ν_e appearance	$\sin^2\theta_{23}, \|\Delta m_{23}^2\|$	[72–76]
Solar Neutrinos	SNO	$^8B\ \nu_e$ flux, $^8B\ \nu_x$ flux	$\sin^2\theta_{12}, \|\Delta m_{12}^2\|$	[77]
	SK	$^8B\ \nu_e$ flux		[77, 78]
	Gallex, SAGE	pp + pep + ^{7}Be, ^{8}B + CNO flux		[31, 32]
	Homestake	^{8}B + pep + CNO ν_e flux		[25, 26]
	Borexino	pp, pep, ^{7}Be, ^{8}B ν_e flux ν_e		[79–82]

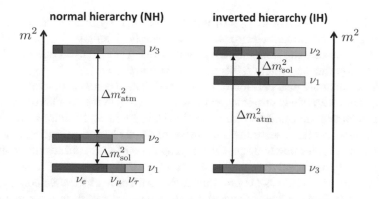

Fig. 1.8 The normal and inverted neutrino mass hierarchies. Uncertainty on the sign of Δm_{23}^2 leads to two neutrino mass orderings which are compatible with current measurements. In the 'normal' ordering, the neutrino state with the largest ν_e component is the lightest. In the 'inverted' ordering the neutrino state with the smallest ν_e component is lightest. From [95]

1.2.5 Mass Scale Measurements

A further unknown is the absolute magnitude of the neutrino masses. The combined measured mass squared differences imply that $\Sigma_{i=0}^3 m_i > 0.05 \pm 0.1\,\mathrm{eV}$ [2], but, other than that, oscillation experiments are not sensitive to the neutrino mass scale. No positive measurement by other means has yet been made, but there are ever more stringent upper bounds being published using cosmological and kinematic constraints.

Kinematic measurements infer the neutrino mass from the energetics of nuclear β^- decays. In these events, a nucleus emits an $\bar{\nu}_e$ and an electron with a total kinetic energy release of:

$$Q_\beta = Q_N - m_e - m_\nu \qquad (1.23)$$

where Q_N is the energy released in the nuclear transition and m_ν, m_e are the neutrino and electron masses, respectively. The large mass of the recoiling nucleus means that it carries away negligible energy. Therefore the maximum electron energy is Q_β, which is determined by the value of m_ν [53, 54]. More generally it can be shown that, close to the end point, a non-zero neutrino mass alters the electron energy spectrum by a constant offset equal to [97]:

$$-m_e^2 = -\sum_{i=0}^{3} |U_{ei}^2| m_i^2 \qquad (1.24)$$

which can be directly measured from the β decay spectrum.

There are three main approaches to this measurement [98]: KATRIN is measuring the energy spectrum of tritium β^- decay using an electromagnetic spectrometer

[99], the Project8 experiment [100] will do the same by measuring the Larmour frequencies of the emitted electrons in a magnetic field, and the ECHO experiment [101] will measure the holmium-63 spectrum using low-temperature calorimetry. All three experiments target $\mathcal{O}(100\,\text{meV})$ sensitivities on m_e.

Indirect cosmological constraints are derived from the effects of a finite neutrino mass on the well verified cosmological standard model. In the ΛCDM cosmological model, neutrinos are created in the big bang. They are in equilibrium with the other particles until cooling causes them to freeze-out,[6] just before the recombination of electron and protons into hydrogen atoms. The mass of the neutrino adds additional matter density to the universe and the exact value of m_ν determines how long neutrinos remain relativistic as the universe cools. Both have observable consequences for large scale structure formation and the power spectrum of the cosmic microwave background. A recent analysis combining many results places an upper limit on the sum of the masses [102]:

$$\sum_{i=0}^{3} m_i < 0.17\text{eV} \tag{1.25}$$

at 95% confidence.

1.3   What Is the Neutrino?

Though the evidence for the neutrino mass is indisputible, there is no single obvious path for extending the SM to incorporate it. The next section outlines each of the options, starting from two foundational principles: Lorentz covariance, which enforces compatibility with special relativity, and gauge invariance, which correctly describes all particle interactions.

To ensure that the neutrino field is Lorentz covariant and has spin-1/2, it must be built from the left and right handed Weyl spinors, the $(0, 1/2)$ and $(1/2, 0)$ representations of the Lorentz group.[7] This leads to three possible descriptions of the neutrino: the Weyl spinor, used to describe the massless neutrino in the SM; the Dirac spinor, used to describe all massive fermions of the SM; and the Majorana spinor, appropriate only for the neutrino. It will become clear that only the latter two can accomodate a neutrino mass.

As each of these descriptions are considered in turn, particles represented by Majorana and Dirac spinors will be referred to as Majorana and Dirac particles respectively.

---

[6] As the universe expands, neutrinos collide with other particles ever less frequently. Eventually, interactions betweeen the neutrinos and the rest of the universe's particles are too infrequent to maintain thermal equilibrium between the two and the neutrinos are thermally isolated. This is 'freeze-out'.

[7] An introduction to Lorentz group representations and spin 1/2 fields is given in [103].

Table 1.2 States described by the left/right Weyl spinors

	χ_L	χ_R
Particle helicity	$-1/2$	$1/2$
Anti-particle helicity	$1/2$	$-1/2$

Weyl Neutrinos

The simplest description of the neutrino uses one of the Weyl representations directly. The Weyl spinors are 2-component complex spinors denoted by $\chi_{L,R}$; corresponding to the $(1/2, 0)$, $(0, 1/2)$ representations of the Lorentz group, respectively [103].

The Weyl action is:

$$S_R = \int d^4x \, i\chi_R^\dagger \sigma^\mu \partial_\mu \chi_R \tag{1.26}$$

$$S_L = \int d^4x \, i\chi_L^\dagger \overline{\sigma}^\mu \partial_\mu \chi_L \tag{1.27}$$

where $\sigma_\mu = (1, \vec{\sigma})$ and $\overline{\sigma}^\mu = (1, -\vec{\sigma})$, for Pauli spin matrices $\vec{\sigma}$ [103]. Extremising the action gives rise to the Weyl equations of motion:

$$\sigma^\mu \partial_\mu \chi_R = 0 \tag{1.28}$$

$$\overline{\sigma}^\mu \partial_\mu \chi_L = 0 \tag{1.29}$$

Multiplying both of these equations by $\partial_0 \pm \vec{\sigma} \cdot \nabla$ reveals the dispersion relation for a Weyl particle: it gives the Klein-Gordon equation with $m = 0$:

$$\partial_\mu \partial^\mu \chi_{L,R} = 0 \tag{1.30}$$

this shows that the Weyl spinor represents a *massless* particle with $E^2 = p^2 c^2$. One can also show that the plane wave solutions to the equations of motion represent particle and anti-particle states with fixed opposite helicity [103], see table 1.2.

χ_L describes massless spin 1/2 particles which always have helicity $-1/2$ (left handed) and anti-particles which always have helicity $1/2$ (right handed). This is precisely the SM description of the neutrino, but Eq. 1.30 means it cannot be used to describe the massive neutrino.

For χ_R, the helicities are reversed: it gives rise to right handed particles and left handed anti-particles. A neutrino field described by χ_R would give rise to a right handed neutrino and a left handed anti-neutrino. These particles would not feel weak interations and are therefore described as 'sterile'.

Dirac Neutrinos

Massive fermions in the SM sit in the more complicated $(0, \frac{1}{2}) \oplus (\frac{1}{2}, 0)$ representation. This representation mixes the left and right Weyl spinors $\chi_{L,R}$ to give rise to a complex, 4-component Dirac spinor denoted ψ. The Dirac action is [103]:

$$\int d^4x \; \bar{\psi}(i\gamma^\mu \partial_\mu - m)\psi \tag{1.31}$$

where $\bar{\psi} = \psi^\dagger \gamma^0$ and γ^μ are the four dimensional gamma matrices, defined by the Clifford algebra [103]. The resulting equation of motion is the Dirac equation:

$$(i\partial_\mu \gamma^\mu - m)\psi = 0 \tag{1.32}$$

Multiplying both sides of the equation by $i\partial_\mu \gamma^\mu + m$ we again recovers the Klein–Gordon equation:

$$\partial_\mu \partial^\mu \psi + m^2 = 0 \tag{1.33}$$

but this time, with a non-zero mass m.

This Dirac equation imposes 4 constraints on the spinor's 4 complex components, leaving 4 degrees of freedom. To investigate these degrees of freedom, its useful to expand the Dirac equation in the chiral representation of γ^μ

$$\gamma^0 = \begin{pmatrix} 0 & 1 \\ 1 & 0 \end{pmatrix}, \quad \gamma^i = \begin{pmatrix} 0 & \sigma^i \\ -\sigma^i & 0 \end{pmatrix} \tag{1.34}$$

and write the Dirac spinor in terms of two 2D spinors u_+, u_-:

$$\psi = \begin{pmatrix} u_+ \\ u_- \end{pmatrix} \tag{1.35}$$

The Dirac equation then reads:

$$\begin{pmatrix} -m & i\partial_\mu \sigma^\mu \\ i\partial_\mu \bar{\sigma}^\mu & -m \end{pmatrix} \begin{pmatrix} u_+ \\ u_- \end{pmatrix} = \begin{pmatrix} u_+ \\ u_- \end{pmatrix} \tag{1.36}$$

$$\implies \partial_\mu \bar{\sigma}^\mu u_+ = mu_- \tag{1.37}$$

$$\partial_\mu \sigma^\mu u_+ = mu_+ \tag{1.38}$$

When $m = 0$, these are exactly the Weyl equations of motion in Eq. 1.29 with $u_+ = \chi_R$ and $u_- = \chi_L$, so a massless Dirac spinor is identical to two Weyl spinors evolving independently. The Dirac equation admits a mass term by introducing a coupling between these constituent Weyl spinors, on the right of Eq. 1.38. One can also see this by writing the mass term in Eq. 1.33 in terms of its left and right handed components:

$$\psi_L = \begin{pmatrix} 0 \\ u_- \end{pmatrix} \qquad \psi_R = \begin{pmatrix} u_+ \\ 0 \end{pmatrix} \tag{1.39}$$

$$\mathcal{L}^D = m_D \overline{\psi}\psi = -m_D(\overline{\psi_L}\psi_R + \overline{\psi_R}\psi_L) \tag{1.40}$$

The degrees of freedom of ψ are those of both χ_L and χ_R: a left handed particle and a right handed anti-particle from χ_L as well as a right handed particle and a left handed anti-particle from χ_R.

The simplest way to introduce a neutrino mass into the SM is to describe the neutrino with a Dirac spinor, bringing it in line with the charged leptons. This would imply the existence of two new, sterile and as yet unobserved, degrees of freedom in the neutrino field: the right handed neutrino, and the left handed anti-neutrino. However, no neutrino particle would be completely sterile because the helicity states of massive particles are mixtures of left and right handed chiral components. Rather, a left handed Dirac neutrino is *mostly* active and a right handed Dirac neutrino is *mostly* sterile.

Majorana Neutrinos

In 1937, Ettore Majorana asked if a mass term could be generated using only the left handed neutrino and right handed anti-neutrino, removing the need to introduce new sterile degrees of freedom [2]. He found that two degrees of freedom could be removed from the Dirac field by imposing the Majorana condition:

$$\psi^{(c)} = \psi \tag{1.41}$$

where the charge conjugate field $\psi^{(c)}$ is defined by:

$$\psi^{(c)} = -\gamma^0 \mathcal{C}\psi^* \tag{1.42}$$

and $\mathcal{C}$ is the charge conjugation operator. Charge conservation demands that the Majorana condition can only hold for neutral particles, because if ψ has charge q, ψ^* has charge $-q$. The only known neutral fermion is the neutrino.

It can be shown that Eq. 1.41 is compatible with the Dirac equation and that it is preserved in Lorentz transformations [2, 104], so the condition is physically meaningful, and the spinor it applies to can represent a massive particle.

A Dirac spinor that satisfies Eq. 1.41 is called a Majorana spinor. Writing Eq. 1.41 in the chiral basis[8] reveals the degrees of freedom of the Majorana spinor:

---

[8]Where $\mathcal{C} = i\gamma^2\gamma^0$.

$$-i\gamma^0 C \psi^* = i\gamma^2 \psi^* = \begin{pmatrix} 0 & i\sigma^2 \\ -i\sigma^2 & 0 \end{pmatrix} \begin{pmatrix} \chi_L^* \\ \chi_R^* \end{pmatrix} = \begin{pmatrix} \chi_L \\ \chi_R \end{pmatrix} \tag{1.43}$$

$$\implies \begin{pmatrix} \chi_L \\ \chi_R \end{pmatrix} = \begin{pmatrix} i\sigma^2 \chi_R^* \\ -i\sigma^2 \chi_L^* \end{pmatrix} \tag{1.44}$$

So, as before, there are two coupled Weyl spinors, but they are not independent: one can be eliminated to give:

$$\psi = \begin{pmatrix} \chi_L \\ -i\sigma^2 \chi_L^* \end{pmatrix} \tag{1.45}$$

This shows that a Majorana spinor has just two degrees of freedom, those of the left handed Weyl spinor and that the mass is introduced via a χ_L self-interaction. Again, this is made clear by writing the mass term in terms of the left and right handed components of the field:

$$\mathcal{L}^M = -m_D(\overline{\psi_L}\psi_R + \overline{\psi_R}\psi_L) = -m_L(\overline{\psi_L^{(c)}}\psi_L + \text{H.c.}) \tag{1.46}$$

m_L here is known as the Majorana mass, H.c. denotes the Hermitian conjugate. As before, the mass is generated by coupling between left and right handed components, but now the right handed component is the dependent conjugate of the left handed component.

When built from left handed fields, this description seems tailor-made for the massive neutrino: it has the correct two degrees of freedom and a non-zero mass. One could also build a Majorana spinor from right handed fields, giving rise to a right handed particle and a left handed anti-particle. Identically to the Dirac case, these fields would not feel weak interactions and are considered sterile.

A critical difference emerges when one considers the mass states of Dirac and Majorana neutrinos. In both cases, the mass states have both left and right chiral components. For Dirac neutrinos, the mixture is between left and right handed, active and sterile. For Majorana neutrinos the picture is radically different: the mixtures contain both ν_L and $\nu_L^{(c)}$, which behave like the SM particle and anti-particle, respectively. This mixing of particle-like and anti-particle-like completely removes the distinction between Majorana particle and Majorana anti-particle.

1.3.1 A Majorana Mass Means Majorana Neutrinos

It has been shown that there are two possible descriptions of the massive neutrino: Dirac and Majorana spinors which give rise to Dirac and Majorana masses, respectively. It is worth asking how these two concepts are related: can a Dirac particle have a Majorana mass and, if yes, is it still a Dirac particle? In fact, a non-zero Majorana mass always implies the neutrino is a Majorana particle.

To see this, write the most general neutrino mass Lagrangian, containing both left and right handed neutrino fields ν_L, ν_R, Majorana masses for both ($m_{L,R}$) and a Dirac mass that couples them (m_D).

$$\mathcal{L}_{mass} = \mathcal{L}_{mass}^{D} + \mathcal{L}_{mass}^{L} + \mathcal{L}_{mass}^{R} \tag{1.47}$$

Written in terms of the left handed chiral fields:

$$N_L = \begin{pmatrix} \nu_L \\ \nu_R^{(c)} \end{pmatrix} = \begin{pmatrix} \nu_L \\ C\overline{\nu_R}^T \end{pmatrix} \tag{1.48}$$

the Lagrangian is:

$$\mathcal{L}_{mass} = \frac{1}{2} N_L^T C^\dagger M N_L + \text{H.c.} \tag{1.49}$$

which defines the neutrino mass matrix M:

$$M = \begin{pmatrix} m_L & m_D \\ m_D & m_R \end{pmatrix} \tag{1.50}$$

Because this matrix is non-diagonal, it implies that ν_L, ν_R do not have definite mass. The fields with definite mass are apparent in the basis where M is diagonal:

$$M^D = \begin{pmatrix} m_1 & 0 \\ 0 & m_2 \end{pmatrix} \qquad N^D = \begin{pmatrix} \nu_{1L} \\ \nu_{2L} \end{pmatrix} \tag{1.51}$$

In this rotated basis, the Lagrangian reads:

$$\mathcal{L}_{mass}^{M+D} = \frac{1}{2} \sum_{k=1,2} m_k \nu_{kL}^T C^\dagger \nu_{kL} + \text{H.c.} \tag{1.52}$$

defining $\nu_k = \nu_{kL} + \nu_{kL}^{(c)}$, this becomes:

$$\mathcal{L}_{mass}^{M+D} = -\frac{1}{2} \sum_{k=1,2} m_k \overline{\nu}_k \nu_k \tag{1.53}$$

This is the expression for two Majorana spinors, each with its own Majorana mass. Therefore, if the neutrino has *any* Majorana mass, its correct description is a Majorana spinor, regardless of any Dirac terms.

1.4   Neutrino Mass Mechanism

A neutrino field described by Dirac or Majorana spinors correctly attributes masses to the neutrino of the following forms:

Table 1.3 Weak charges for the electron and neutrino fields

Field	I_3	Y
e_r	0	-2
e_l	$-1/2$	-1
ν_l	$1/2$	-1

$$-m_D(\bar{\psi}_L\psi_R + \bar{\psi}_R\psi_L) \tag{1.54}$$

$$-m_L(\overline{\psi}_L^{(c)}\psi_L + \text{H.c.}) \tag{1.55}$$

Unfortunately though, *both* of these terms are forbidden by the foundational principle which describes interactions between particle fields: gauge invariance. Before the development of electro-weak symmetry breaking theory, the indisputable evidence for the mass of all fermions directly contradicted the theory which correctly predicted their interactions. The theory of the Higgs field and the subsequent discovery of the Higgs boson solved this problem for the massive fermions and vector bosons of the SM, but the neutrino mass still evades the same certainty. The next section briefly outlines gauge invariance in the electroweak sector, its conflict with the neutrino mass, and the theoretical mechanisms that hint at a solution.

1.4.1   Gauge Invariance

The electroweak sector of the SM requires that physics is invariant under local transformations belonging to the symmetry group $SU(2)_L \times U(1)_Y$.[9] To ensure this, each SM field is placed in a gauge multiplet, a combination of fields that transforms according to a particular representation of the gauge group. The charge of each particle under electroweak interactions is determined by the choice of representation, which can be selected to match experiment. These charges are the are weak isospin, I_3 and weak hyper-charge, Y, shown for the electron and neutrino in Table 1.3. The symmetry also *forces* the introduction of four new vector fields that represent the force carriers of the electroweak interaction. Permitting all gauge invariant combinations of these fields to the SM Lagrangian and no others successfully describes all electroweak phenomena.

However, a problem arises when considering the charge of the Dirac mass term in Eq. 1.55 for the electron. e_L and e_R have different I_3 and therefore the two contributions to the Dirac mass behave differently under gauge transformations and their sum cannot be gauge invariant. If the neutrino was promoted to a Dirac particle, it would create exactly the same problem. Furthermore, the Majorana mass term carries a weak hyper-charge of $-1 + -1 = -2$. A non-zero charge means that the mass term changes under gauge transformations and cannot be gauge invariant.

---

[9]References [2, 105] serve as a good introductions for the following two sections.

1.4.2 Higgs to the Rescue

Is it possible to reconcile the success of the $SU(2)_L \times U(1)_Y$ model with the fact we don't live in a massless universe? For Dirac particles, the solution is to promote the mass to dynamical field that carries electroweak charges: the Higgs field.

The Higgs field is a Lorentz scalar complex $SU(2)$ doublet:

$$\Phi = \begin{pmatrix} \Phi^+ \\ \Phi^0 \end{pmatrix} \tag{1.56}$$

where both components carry $Y = 1$, and Φ^+, Φ^0 carry I_3 of $1/2$, $-1/2$ respectively. If the left handed electron and neutrino fields are written together in an $SU(2)$ doublet:

$$L_L = \begin{pmatrix} \nu_L \\ e_L \end{pmatrix} \tag{1.57}$$

it is possible to couple the Higgs to the fermion fields, with interaction strength g, forming a term that has no overall charge, making it gauge invariant:

$$- g \left(\bar{\nu}_L \ \bar{e}_L \right) \begin{pmatrix} \Phi^+ \\ \Phi^0 \end{pmatrix} e_R + \text{H.c.} \tag{1.58}$$

The significance of this term becomes apparent because, below a certain temperature, the Higgs field obtains a vacuum expectation value (V.E.V):

$$< \Phi > = \begin{pmatrix} 0 \\ \frac{v}{\sqrt{2}} \end{pmatrix} \tag{1.59}$$

after this, the Higgs coupling includes terms of the form:

$$- \frac{gv}{\sqrt{2}} (\overline{e_L} e_R + \overline{e_R} e_L) \tag{1.60}$$

which is just the mass term in Eq. 1.55, with

$$m = \frac{gv}{\sqrt{2}} \tag{1.61}$$

The Higgs mechanism allows the theory to have its cake and eat it too: the electron field is massive and its Lagrangian is gauge invariant.

If there is a right handed neutrino field, ν_r, it is possible to generate a Dirac mass for the neutrino using the Higgs mechanism in the same way. The coupling is:

$$\mathcal{L}_{mass} = -g_\nu (\overline{\nu_R} \tilde{\Phi}^\dagger L_L + \overline{L_L} \tilde{\Phi} \nu_R) \tag{1.62}$$

Fig. 1.9 Standard model
particle masses. The neutrino
ranges show the masses
which are compatible with
current experimental results

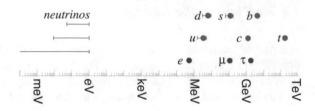

where

$$\tilde{\Phi} = i\tau_2 \Phi^* \gamma^0 \tag{1.63}$$

and τ_2 is the Pauli matrix. After the Higgs obtains a V.E.V, this term produces an mass term for the neutrino identical to the one in Eq. 1.61 but with coupling g_ν.

One concern with this solution is the 'unnaturally' small Yukawa coupling, g_v, that would be required to describe a neutrino mass of 170 meV [2].[10] Figure 1.9 shows the masses of the SM particles on a log-scale; the neutrino mass is at least 6 orders of magnitude smaller than the next lightest, the electron. This requires that the Higgs—neutrino coupling is also 6 orders smaller, which some consider unlikely.

It is natural to ask if the Higgs field can also make a Majorana mass gauge invariant. The Majorana mass term carries a charge of $Y = \pm 2$, so two copies of the Higgs field would be required. It is simple to show that such a term is non-renormalisable.[11] This restriction means that the SM Lagrangian cannot contain a true Majorana mass m_L. However, m_L can emerge as a low energy approximation to a BSM theory and produce the same phenomenology.

1.4.3 Effective Field Theory

Fortunately, there is reason to expect an emergent m_L. Many BSM theories treat the SM as an effective low-energy theory resulting from the spontaneous symmetry breaking of a larger symmetry group. Then, the leading order corrections to the SM are non-renormalisable, effective low-energy Lagrangian terms, built from SM fields and respecting SM symmetries. The highest order correction satisfying these constraints involves the neutrino field [2]:

$$\mathcal{L}_5 = \frac{g}{\mathcal{M}}(L_L^T \tau_2 \Phi) C^\dagger (\Phi^T \tau_2 L_L) + \text{H.c.} \tag{1.64}$$

where $\mathcal{M}$ is the scale of new physics that generates the term and L_L is the lepton SU(2) doublet. When the Higgs field obtains a V.E.V., this term generates a Majorana mass term for ν_L:

---

[10]The largest mass currently allowed by cosmology.

[11]Fermion fields carry dimension $[E]^{\frac{3}{2}}$, so the Majorana mass term carries dimension $[E]^3$ whilst boson fields carry dimension $[E]$. Two Higgs fields in the interaction term would give it dimension 5, Lagrangian terms of dimension greater than 4 are non-renormalisable.

$$< \Phi > \to \begin{pmatrix} 0 \\ v \end{pmatrix} \qquad \mathcal{L}_5 = \frac{1}{2}\frac{gv^2}{\mathcal{M}}\overline{\nu_L^{(c)}}\nu_L \tag{1.65}$$

with mass:

$$m_\nu = \frac{gv^2}{\mathcal{M}} \tag{1.66}$$

therefore Majorana neutrino masses are expected to appear in many BSM theories, unless some other symmetry forbids them. The simplest theory that generates such a mass supposes that there are right handed neutrino fields ν_R with large Majorana masses, which are not forbidden by gauge invariance.

Note that the Majorana mass decreases as the scale of new physics increases, a feature known as the see-saw mechanism. It is significant for two reasons: first, a large $\mathcal{M}$ can suppress the neutrino mass without requiring an unnaturally small Yukawa coupling, g. Second, it means that searches for the tiny neutrino mass are, in fact, probes of new physics at the high energy scale $\mathcal{M}$.

The effective field theory approach has now provided the final link required to build a gauge invariant description of the massive neutrino. The next section describes one of the most compelling reasons to look for Majorana neutrinos: their possible role in forming today's matter dominated universe.

1.5 Leptogenisis

The Majorana neutrino is related to one of the most profound questions in modern science: what produced the universe's matter/anti-matter asymmetry? Cosmological observations confirm the hot big bang theory, in which there were equal quantities of matter and anti-matter in the early universe [106], but there is almost no anti-matter in today's universe. If at one stage there were equal components of baryonic matter and anti-matter, without an asymmetry in the laws of physics, eventually there would be exact baryon annihilation to γ. This is *almost* what is observed. Our universe is full of photons but with a tiny contamination of baryons. The Planck collaboration recently observed [107]:

$$\frac{n_B}{n_\gamma} = 6.1^{+0.3}_{-0.2} \times 10^{-10} \tag{1.67}$$

which requires an accumulation of baryons over anti-baryons at the level of 1 part per billion before annihilation.

1.5.1 Not Enough CP violation

In 1967, Sakarov published three conditions for baryongenisis, the accumulation of baryons over anti-baryons [108]:

1. Baryon number ($\mathcal{B}$) violation

2. Charge (C) and Charge-Parity (CP) violation
3. Departure from thermal equilibrium

The first condition is self-evident, the third is provided by the expansion of the universe but the second has proved a sticking point. CP violation is observed in the quark sector [35], but the effects are not large enough to explain the observed Baryon asymmetry.

1.5.2 Majorana Phases

A potential solution comes from the PMNS matrix for Majorana neutrinos. To see why, note that the Majorana mass term is not invariant under rephasing of the neutrino fields $\nu_L \rightarrow e^{i\theta}\nu_L$, which removes the phase freedom used to eliminate two phases in the PMNS matrix [20]. The result is two additional Majorana phases $\phi_{1,2}$ which become physical for Majorana neutrinos. The Majorana and Dirac PMNS matrices $U_{M,D}$ are related by:

$$U_M = U_D \cdot P_M \qquad P_M = \text{diag}(1, e^{i\phi_1}, e^{i\phi_2}) \tag{1.68}$$

Like δ_{CP}, $\phi_{1,2}$ lead to CP violation, perhaps in the quantities Sakarhrov required.

1.5.3 Heavy Neutrino Decay

Fukugita et al. [109] proposed a mechanism for realising baryogenisis using $\phi_{1,2}$. If the heavy neutrino states described in Sect. 1.4.3 do exist, they would have been created in the big bang. As the universe cooled, these heavy neutrinos would decay via their Yukawa couplings to Higgs particles $\overset{(-)}{\phi}$ and charged leptons $\overset{(-)}{l}$, as depicted in Fig. 1.10.

The Majorana phases lead to CP violation in these decays, so that the rate of decay to leptons is not equal to the rate of decay to anti-leptons:

$$R(N_R \rightarrow \phi + \bar{l}) \neq R(N_R \rightarrow \bar{\phi} + l) \tag{1.69}$$

this would produce a $l, \bar{l}$ asymmetry and an accumulation of matter over anti-matter [109] called leptogenisis. It is hypothesised that any asymmetry could be transferred to the baryon sector via so called 'sphaleron' processes that convert baryons to anti-leptons and anti-baryons to leptons [110].

Therefore, if the neutrino is a Majorana particle, the decay of right handed particles in the early universe could fuel the unexplained matter/anti-matter asymmetry observed today.

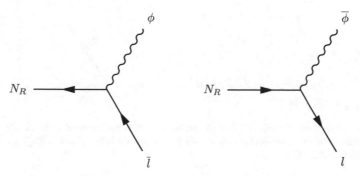

Fig. 1.10 Heavy N_R decay to (anti-)leptons, $\overset{(-)}{l}$, and Higgs particles, $\overset{(-)}{\phi}$ which could have driven leptogenisis in the early universe. Diagrams adapted from [109]

1.6 Observing Majorana Neutrinos

There are several compelling reasons to look for Majorana neutrinos. First, Occam's razor says that they're there: the success of the SM rests on the principle of including all gauge invariant terms in the Lagrangian, unless some symmetry explicitly forbids them. No symmetry prevents a heavy right handed Majorana mass, m_R, which would make the neutrino Majorana, regardless of any Dirac mass. Second, the leading order correction to the SM in effective field theory generates a Majorana mass for the neutrino, making it a Majorana particle. Third, the see-saw mechanism could explain the tiny neutrino mass without resorting to a highly tuned Yukawa coupling. Finally, the decay of heavy Majorana neutrinos in the early universe offers a promising explanation for the universe's matter anti-matter asymmetry. The following section discusses how Majorana neutrinos could be observed.

1.6.1 Flavour Oscillations

A natural place to look for Majorana neutrinos is flavour oscillations, because these are driven by the PMNS matrix, which is different for Dirac and Majorana neutrinos. In particular, the PMNS matrix has two additonal Majorana phases $\phi_{1,2}$. However, there is a well known argument [20, 111] that shows that these phases can have no effect on oscillations. Equation 1.11 showed that the flavour oscillation probability depends only on the combination $UTU^\dagger$, where T is the mass basis representation of the time evolution operator (Eq. 1.12) and U is the PMNS matrix. This combination is equal for Dirac and Majorana neutrinos:

$$U_M T U_M{}^\dagger = U_D P_M T P_M^\dagger U_D{}^\dagger = U_D T U_D{}^\dagger \qquad (1.70)$$

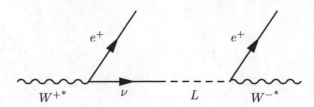

Fig. 1.11 A hypothetical $\nu_e \rightarrow \bar{\nu}_e$ Majorana neutrino experiment. The production of two same-sign leptons is a signature for lepton number violation. The process is possible for Majorana neutrinos but not Dirac neutrinos

The final step relies on the fact that both T and P_M are diagonal in the mass basis and therefore commute. This means there can be no observable differences in neutrino oscillation phenomena between Dirac and Majorana neutrinos. This argument can be extended to show the same result for oscillations in matter and for additional numbers of neutrino fields [112].

1.6.2 Lepton Number Violation

A second, more promising, possibility comes from the Majorana mass term. After second quantisation, the term $\overline{\psi_L^{(c)}}\psi_L$ leads to a vertex that creates two (anti-)neutrinos, changing the overall lepton number by ±2. Clearly, lepton number is no longer a good quantum number in a universe with Majorana neutrinos, whereas lepton number is a conserved quantity in the SM.

Equivalently, Sect. 1.3 showed that there is no distinction between particle and anti-particle for Majorana spinors. Therefore, Majorana neutrinos produce processes in which a SM ν-like particle later behaves like a SM $\bar{\nu}$. Figure 1.11 shows a cartoon of an experiment to look for this behaviour. It produces ν_e and then looks for $\bar{\nu}_e$-like creation of e^+ some time later. Observing e^+ is a positive signal for Majorana neutrinos. To see why, it is useful to consider the helicity of the intermediate neutrino. For both Dirac and Majorana neutrinos, the ν_e created in the first vertex almost always has left handed helicity, because the weak interaction couples only to the left handed chiral part of the field. However, differences emerge between the two in the rare case that the ν_e is created with right handed helicity. In the Dirac case, the right handed ν_e is almost completely sterile and will not interact at all but, for a Majorana neutrino, the right handed $\overset{(-)}{\nu}_e$ will behave as if it were a SM $\bar{\nu}_e$, possibly creating an e^+.

Though this experiment works in principle, the problem is that the creation of the right handed helicity neutrino is suppressed[12] by a factor of $(m_\nu/E_\nu)^2$, for both Majorana and Dirac neutrinos. Given the produced neutrinos must have at least $0.5\,\mathrm{MeV}$ to create the second charged lepton and the neutrino mass is $<1\,\mathrm{eV}$, this is at most:

---

[12]This is just the same helicity suppression that causes $\pi^\pm$ to decay predominantly to μ.

$$\left(\frac{m_1}{E}\right)^2 < \left(\frac{1\,\text{eV}}{500\,\text{keV}}\right)^2 = 4 \times 10^{-12} \tag{1.71}$$

which requires a completely infeasible neutrino flux and background rejection.

The next section describes neutrinoless double beta decay, an analogous nuclear decay process. There, helicity suppression is overcome with Avagadro's number, by amassing a very large number of nuclei.

1.7 Neutrinoless Double Beta Decay

In this final section outlines the only viable experimental signature for Majorana neutrinos: the possibility of neutrinoless double beta decay ($0\nu\beta\beta$). It describes contemporary searches for it and introduces the SNO+ experiment, the focus of this work.

1.7.1 Double Beta Decay

Figure 1.12 is a cartoon showing the trend in atomic mass isobars for even A nuclei. The odd-Z, odd-N nuclei sit higher in energy than the even-Z, even-N nuclei because paired nucleons are more strongly bound.[13]

$\beta^\pm$ decay occurs when a nucleus can lower its energy by altering its Z by ± 1, moving left/right across the diagram by one unit. A proton is converted into a neutron (or vice-versa), emitting an electron and a neutrino:

$$(Z, A) \rightarrow (Z \pm 1, A) + e^\mp + \overset{(-)}{\nu_e} \tag{1.72}$$

For some nuclei, e.g. the one at $Z - 2$ in Fig. 1.12, $\beta^\pm$ decay would *raise* the energy and is forbidden. However, some of these nuclei can transition to a lower energy by simultaneously undergoing two $\beta^\pm$ decays. This is double beta decay ($2\nu\beta\beta$), first proposed M. Goeppert-Mayer in 1935 [114]. During a double beta decay, the nucleus emits two electrons and two anti-neutrinos, changing its Z by two units:

$$(Z, A) \rightarrow (Z \pm 2, A) + 2\beta^\mp + 2\overset{(-)}{\nu_e} \tag{1.73}$$

and releasing total kinetic energy $Q_{\beta\beta}$[14]:

$$Q_{\beta\beta}/c^2 = m(Z, A) - m(Z, A \pm 2) - 2m_e \tag{1.74}$$

---

[13]This is the 'pairing' term of the semi-empirical mass formula for nuclei.
[14]Neglecting the neutrino mass.

Fig. 1.12 β and $\beta\beta$
transitions for even A nuclei.
The blue lines show the trend
in nuclear energy (isobars);
the black dots show example
nuclear energies. For some
even-Z, even-N nuclei β
decay is forbidden, but
double beta decay ($\beta\beta$) is
allowed. From [113]

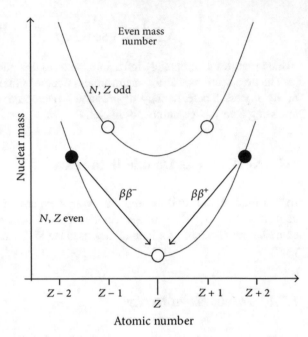

Fig. 1.13 Quark-level
Feynman diagram for $2\nu\beta\beta$

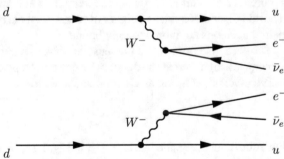

Figure 1.13 shows the quark level Feynman diagram for $2\nu\beta\beta$. It is a second order
weak process which will be sub-dominant to single β decay when the latter is kine-
matically allowed.

There are 35 known naturally occurring nuclei that could undergo $2\nu\beta\beta$ [35, 115]
but, in practice, the decay can only be observed in those isotopes for which single β
and α decay are forbidden or strongly suppressed.[15] Like the hypothetical example
in Fig. 1.12, those nuclei are all even-even $\rightarrow$ even-even transitions.

First directly observed in ^{82}Se in 1987 by Elliot, Hahn and Moe [117], there
are now 20 measured double beta isotopes [35]. ^{130}Te, which is the focus of the
SNO+ experiment, has a measured $2\nu\beta\beta$ half-life of 8.2 ± 0.2(stat) ± 0.6(sys) $\times$
10^{20}yr [118].

---

[15]The exception is the α emitter ^{238}U for which the half-life was measured radio-chemically [116].

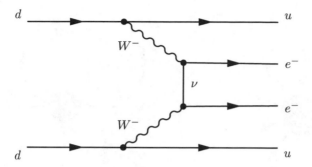

Fig. 1.14 Quark-level Feynman diagram for $0\nu\beta\beta$ via light neutrino exchange

1.7.2 Neutrinoless Decay

In 1948, Furry proposed that double beta decay could proceed without the emission of neutrinos, leaving just two electrons in the final state [119, 120]:

$$(Z, A) \rightarrow (Z, A \pm 2) + 2\,\beta^{\mp} + 0\,\overset{(-)}{\nu}_e \tag{1.75}$$

Clearly this process violates lepton number by $\Delta L = \pm 2$, and cannot occur in the SM. However, if neutrinos are massive Majorana particles, neutrinoless double beta decay ($0\nu\beta\beta$) is possible, if mediated by a virtual Majorana neutrino, as shown in Fig. 1.14.

In fact, the converse is also true: the existence of $0\nu\beta\beta$ decay implies that neutrinos are Majorana particles, regardless of the mechanism. This is proven by the black-box theorem [121, 122], depicted in Fig. 1.15. The argument states that if $0\nu\beta\beta$ is possible, one can always draw Feynman diagram whose overall effect is $2d \rightarrow 2u + 2e^-$ with an unknown vertex (left). This vertex can always be re-arranged to create $\nu \rightarrow \bar{\nu}$ transitions (right), that generate a Majorana mass term of the form given in Eq. 1.55. This Majorana mass, in turn, implies that the neutrino is a Majorana particle, by the argument given in Sect. 1.3.1.

The black box theorem is important because the Feynman diagram shown in Fig. 1.14 is only the most popular of many possible $0\nu\beta\beta$ mechanisms predicted by BSM theories. Alternatives emerge in extra dimension theories, models with leptoquarks, $\mathcal{R}$ parity violating super-symmetry, and left-right symmetric models [123]. Indeed, several mechanisms may contribute and interfere.

Of particular importance in this work are the right-handed currents which emerge in left-right symmetric theories. These theories posit that the electro-weak interaction is left-right symmetric at high energies. This symmetry is spontaneously broken by a Higgs field, at a temperature above the electro-weak symmetry breaking scale, to produce massive right-handed gauge bosons, massive right-handed neutrinos and an expanded Higgs sector. The interactions of these particles give rise to several

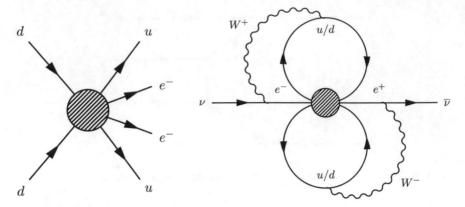

Fig. 1.15 Quark-level Feynamn diagrams for the black box theorem. Left: a general $0\nu\beta\beta$ mechanism. Right: the resulting Majorana mass term generated. The shaded region shows the overall effect of new physics

Fig. 1.16 Feynman diagram for $0\nu\beta\beta$ via heavy neutrino exchange with right-handed currents

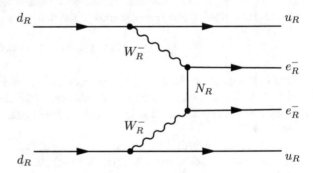

alternative $0\nu\beta\beta$ diagrams. In the simplest models, the right-handed current diagram shown in Fig. 1.16 is expected to dominate [123].

1.7.3 $0\nu\beta\beta$ Rates

The rate of neutrinoless double beta decay depends on both the nuclear physics of the transition, and the particle physics in Fig. 1.14. For the light neutrino exchange mechanism, the particle physics contributes a term proportional to the effective Majorana mass of the neutrino $m_{\beta\beta}$:

$$m_{\beta\beta} = \sum_{k=0}^{3} U_{ek}^2 m_k \tag{1.76}$$

which can be easily read off from Fig. 1.14. The two factors of U_{ek} come from the fact that the weak interaction projects out only the ν_e component of the neutrino, the

Fig. 1.17 $m_{\beta\beta}$ as a function of lightest neutrino mass [125]. IH/NH indicate the inverted/normal mass hierarchies. m_{min} is the (partly constrained) lightest neutrino mass; $m_{\beta\beta}$ is the effective Majorana mass. The bands represent increasing uncertainty in the neutrino mixing angles and mass splittings, as well as the (unknown) Majorana phases

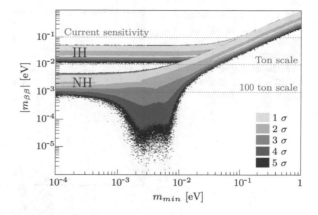

sum over mass states k comes from that fact that any of the three massive neutrino states could be exchanged, and m_k comes from helicity suppression: one of vertices must create a left handed ν, so that it may be absorbed at the other. This suppression means $0\nu\beta\beta$ will have far lower rates than $2\nu\beta\beta$ for realistic values of m_k.

The nuclear and particle physics contributions are separable to good approximation, so the half-life may be written as the product of the nuclear matrix element $M^{0\nu}$, the phase space of the decay $G^{0\nu}$ and $m_{\beta\beta}$ normalised to m_e [124]:

$$\frac{1}{T_{1/2}} = G^{0\nu}||M^{0\nu}||^2 \left(\frac{|m_{\beta\beta}|}{m_e}\right)^2 \tag{1.77}$$

$m_{\beta\beta}$, as stated in Eq. 1.76, hides a lot of complexity. It depends on all three of the absolute neutrino masses and the elements of the PMNS matrix, including the unknown Majorana phases $\phi_{1,2}$. Figure 1.17 shows $m_{\beta\beta}$ as a function of the lightest neutrino mass. The shaded region shows the possible values of $m_{\beta\beta}$, allowing for $\phi_{1,2} \in [0, 2\pi]$ and 1–5σ variation in the measured mass splittings and mixing angles. There is a strong dependence on the mass hierarchy below a degenerate regime,[16] $m_{min} < 0.1$ eV, where $0\nu\beta\beta$ rates are 2–4 orders of magnitude lower for the normal hierarchy. There is also the possibility of complete cancellation between terms in Eq. 1.76 for the normal hierarchy and a lightest neutrino mass of around 4 meV.

The Feynman diagram for $0\nu\beta\beta$ is related to the hypothetical Majorana neutrino experiment discussed in Sect. 1.11 by a simple rotation, so it's worth asking what makes this experiment any more feasible. The answer is that it is possible to amass many more nuclei of $2\nu\beta\beta$ isotope than the number of ν that could be produced in a collider, and, rather than wrong sign beam contamination, the dominant background is now low energy radioactivity which can be suppressed using many established techniques.

---

[16]Part of which is not ruled out by cosmology.

Fig. 1.18 A cartoon of the visible kinetic energy spectra of $2\nu\beta\beta$ and $0\nu\beta\beta$. The signature in any real experiment will be smeared by finite energy resolution. Courtesy of E. Fiorini

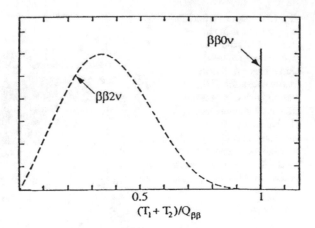

$$\frac{(T_1 + T_2)}{Q_{\beta\beta}}$$

Unfortunately, introducing nuclear physics comes with two costs. First, the quantity of interest, $m_{\beta\beta}$, is proportional to the square root of the observable $T_{1/2}$ which leads to poor sensitivity scalings. Second, there is considerable uncertainty in the values of $M^{0\nu}$, which limits mass measurements and makes comparison between isotopes difficult. A recent review of the field in [124] shows that there is a factor of three uncertainty in $M^{0\nu}$ for ^{130}Te.

KamLAND-Zen currently holds the most stringent limit on $m_{\beta\beta}$, measured with ^{136}Xe [126]:

$$m_{\beta\beta} < 61 - 165\,\text{meV at 90\% confidence} \tag{1.78}$$

where the range reflects uncertainty on the nuclear matrix element of ^{136}Xe. Most of the $m_{\beta\beta}$ range allowed by cosmology is unexplored.

1.7.4   Experimental Signature

$0\nu\beta\beta$ events may be identified using the characteristic energy spectrum they produce in a detector. $2\nu\beta\beta$ and $0\nu\beta\beta$ produce the same amount of kinetic energy,[17] $Q_{\beta\beta}$, but they differ in the number of particles in the final state. The recoil energy of the heavy nucleus is negligible, so, in $2\nu\beta\beta$, $Q_{\beta\beta}$ is shared between four light particles. The neutrinos typically escape detection, so the total visible energy from the electrons has a continuous spectrum with an end point at $Q_{\beta\beta}$, see Fig. 1.18. However, in $0\nu\beta\beta$, the electrons take all of the energy and the visible energy is fixed at the end point of the $2\nu\beta\beta$ spectrum. Therefore, the experimental signature is an excess of two-electron like events at $Q_{\beta\beta}$.

$0\nu\beta\beta$ is a three body decay, so there is no kinematic constraint on the separation or energy split between the two electrons. Rather, the distributions of these two

---

[17]Ignoring the neutrino mass.

Fig. 1.19 $0\nu\beta\beta$ decay kinematics for the light neutrino exchange (LNE) and right handed current (RHC) models. Top: the cosine of the angle separating the emitted electrons (MC truth). Bottom: the fraction of the total energy release taken by electron 1 (arbitrarily chosen). Generated using RAT 6.1.6, which implements the distributions in [127]

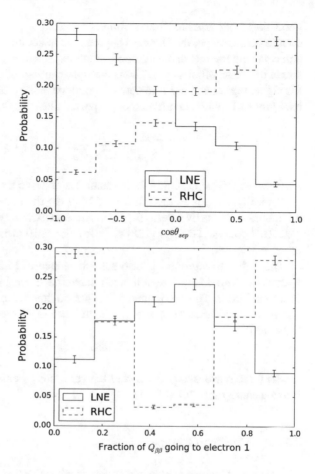

quantities depend strongly on the underlying particle physics mechanism. Figure 1.19 compares the cosine of the angle separating the electrons and the fraction of the energy going to the first electron for the light neutrino exchange and the right handed current mechanisms [127]. In general, light neutrino exchange events tend to emit the electrons back-to-back, with similar energy. Conversely, the right handed current events tend to emit the electrons co-linearly, with one electron taking most of the energy.

1.7.5 $0\nu\beta\beta$ Experiments

At the maximum allowed $m_{\beta\beta}$ of around 100 meV, SNO+ would expect just 2.5 counts/100 kg/yr of $0\nu\beta\beta$ in natural Te[18]. In these circumstances, what makes a

---

[18] Assuming $G^{0\nu} = 3.96 \times 10^{-14}$, $M^{0\nu} = 4.03$.

good $0\nu\beta\beta$ experiment? For a simple counting experiment using $N_{\beta\beta}$ nuclei of double beta isotope, the standard approach is to set up a signal region around $Q_{\beta\beta}$ and constrain the rate of background counts expected, r_B, during live-time, t, and the signal detection efficiency, ϵ. Then, count the number of events actually observed in the signal region and use it to derive a confidence limit on the decay half-life. For a background limited experiment, the expected classical 90% confidence limit is[19]:

$$T_{\frac{1}{2}}^{90\%} = \frac{\ln 2}{1.28\sqrt{r_B}} \cdot \epsilon \cdot \sqrt{t} \cdot N_{\beta\beta} \qquad (1.79)$$

so, $0\nu\beta\beta$ experiments require large quantities of $2\nu\beta\beta$ isotope and low backgrounds.

Typically, $Q_{\beta\beta}$ is in the range $0-5$ MeV, so the background can be dominated by natural radioactive decays [115]. Air and water typically have activities of $\approx 3 \times 10^{10}$ counts/100 kg/yr [128] of ^{222}Rn alone, 10 orders of magnitude above the expected signal.

The $2\nu\beta\beta$ background is also a major concern. Finite energy resolution both widens the $0\nu\beta\beta$ signal region and smears the much larger, steeply falling $2\nu\beta\beta$ spectrum into it. This compound effect means that the expected $2\nu\beta\beta$ counts ($B_{2\nu}$) a is non-linear function of energy resolution. For Gaussian smearing of width σ_E [129, 130]:

$$B_{2\nu} \sim \sigma_E^{5.5} \qquad (1.80)$$

For these reasons, $0\nu\beta\beta$ searches tend to focus on extreme radio-purity and minimising energy resolution [131].

1.7.5.1   Choice of Isotope

The choice of isotope is also key to determining the size of the $0\nu\beta\beta$ signal and the expected backgrounds.

There are three considerations in determining N_I: the procurement cost, the natural abundance and the enrichment cost. A high isotope abundance means greater $N_{\beta\beta}$, without the expense, engineering challenge and additional background of a larger detector. Typically isotope enrichment is expensive $> 10\,\$g^{-1}$ [130], so a reasonable natural abundance is critical.

The nuclear physics of the isotope also determines the expected signal size. Isotopes with larger $0\nu\beta\beta$ nuclear matrix elements and phase space factors produce larger $0\nu\beta\beta$ signals for the same $m_{\beta\beta}$. In practice, the ratios $\frac{||M^{0\nu}||}{||M^{2\nu}||}$ $\frac{G^{0\nu}}{G^{2\nu}}$ are also important, given any search can be limited by the $2\nu\beta\beta$ mis-reconstructions.

The position of the end point, $Q_{\beta\beta}$, has three major impacts. First, larger energy releases are associated with larger phase space factors. Second, a larger energy release will be easier to detect with high efficiency. Third, if the end point lies many res-

---

[19]In the limit that the number of background counts is Gaussian.

Table 1.4 Summary of low background $0\nu\beta\beta$ experiments

Experiment	Isotope	Isotope Mass/kg	Technology	FWHM energy resolution at $Q_{\beta\beta}$	Particle ID	Latest limit (90% confidence)	Reference
CUORE	^{130}Te	206	Crystal bolometer	7.7 ± 0.5 keV	–	1.5×10^{25}	[132]
GERDA	^{76}Ge	35	Segmented germanium detectors	2.7 keV	Pulse shape discrimination	8.0×10^{25}	[133, 134]
EXO-200	^{136}Xe	160	Time projection chamber	35	Single/multi-site discrimination	1.1×10^{25}	[135]
	^{100}Mo	6.9					[136]
	^{82}Se	0.9				4.6×10^{23}	[136]
	^{116}Cd	0.4				1.0×10^{23}	[137]
NEMO-3	^{130}Te	0.5	Tracking calorimeter	108 (at 3 MeV)	Full tracking	5.0×10^{21}	[138]
	^{150}Nd	37 (g)				5.0×10^{22}	[139]
	^{96}Zr	9				9.2×10^{21}	[140]
	^{48}Ca	7 (g)				2.0×10^{22}	[141]

olution widths away from any natural radioactivity, $0\nu\beta\beta$ events will be easy to discriminate from the radioactivity that is impossible to completely remove from detector materials.

1.7.5.2   Technology

Many orders of magnitude in $m_{\beta\beta}$ are still allowed by experiment so, to stand a good chance of discovering $0\nu\beta\beta$, the sensitivity of experiments must scale well with exposure. Exposure may be increased by extending the live-time or increasing the detector mass. Unfortunately, Eq. 1.79 reveals poor scalings for both: the half-life sensitivity only increases with the square root of the live-time and, if the dominant background scales in proportion to the amount of $2\nu\beta\beta$ isotope ($r_B \propto N_{\beta\beta}$), the sensitivity only increases with the square root of the amount of isotope. Worse still, the measured half-life is only related to the quantity of interest by a square root, $m_{\beta\beta} \propto T_{1/2}^{-1/2}$ and so, overall, the sensitivity scales according to $(t \cdot N_{\beta\beta})^{-1/4}$. $0\nu\beta\beta$ experiments fall into two broad categories, divided by their approach to dealing with this terrible scaling law.

The first approach is to aim for an experiment with close to zero background, using very fine energy resolution and/or particle ID techniques that can reject background. If achieved, the half-life sensitivity increases linearly with exposure and $m_{\beta\beta} \propto (tN_{\beta\beta})^{1/2}$, albeit at the cost of densely instrumented detectors that are often difficult to scale in $N_{\beta\beta}$. A non-exhaustive list of such experiments is given in Table 1.4. The next generation of these experiments targets the inverted hierarchy regime of $m_{\beta\beta}$. Only NEMO-3, and its successor SUPERNEMO [142], are sensitive to the underlying $0\nu\beta\beta$ mechanism via individual tracking of the two emitted electrons.

The alternative approach is to aim for a large and easily scaled $N_{\beta\beta}$, even at the cost of accepting some background. If the dominant component to this background does not scale with $N_{\beta\beta}$, one can still achieve sensitivity gains $\propto N_{\beta\beta}^{-\frac{1}{2}}$, though improvements with live-time are only $t^{-\frac{1}{4}}$.

This approach was pioneered by the KamLAND-Zen experiment [126]. The detector uses Xe-loaded liquid scintillator, containing 340 kg of ^{136}Xe in a balloon placed inside the KamLAND detector. The energy resolution is much wider than the low-background experiments (270 keV FWHM), so their energy region of interest contains significant contamination from $2\nu\beta\beta$ and ^{214}Bi originating on the balloon, but the isotope mass can be scaled up by increasing the enrichment or Xe doping.

The SNO+ experiment, the focus of this work, is pursuing the same background limited, large $2\nu\beta\beta$ mass approach. The SNO detector will be filled with liquid scintillator and loaded with $2\nu\beta\beta$ isotope [143]. ^{130}Te is the chosen isotope: at 34%, it has by far the largest natural abundance of any $2\nu\beta\beta$ emitter, after ^{130}Xe it has the second best expected ratio of $2\nu\beta\beta$ to $0\nu\beta\beta$ [130] and it is transparent in the wavelength region detectable by the SNO photo-multiplier tubes (PMTs) [144]. The Q-value of ^{130}Te is 2.5 MeV; in this region, the dominant radioactivity is $^{212/214}$Bi, which may be rejected using the delayed coincidence technique (Chap. 4). Favourable scaling with isotope mass will be achieved by increasing the loading fraction of Te until the $2\nu\beta\beta$ background dominates. The experiment is described in detail in the following chapter.

References

1. Jensen C, Aaserud F (2000) Controversy and consensus: nuclear beta decay 1911–1934
2. Giunti C, Kim CW (2007) Fundamentals of neutrino physics and astrophysics
3. Brown LM (1978) The idea of the neutrino. Phys Today 31:23–28
4. Pauli W, Pauli letter collection: letter to Lise Meitner. https://cds.cern.ch/record/83282
5. Fermi E (1934) Versuch einer Theorie der β-Strahlen. I. Zeitschrift fur Physik 88:161–177. https://doi.org/10.1007/BF01351864
6. Fermi E (1934) Sopra lo Spostamento per Pressione delle Righe Elevate delle Serie Spettrali, Il Nuovo Cimento (1924-1942) 11(3):157. https://doi.org/10.1007/BF02959829
7. Reines F, Cowan CL (1956) The neutrino. Nature 178:446 EP. https://doi.org/10.1038/178446a0
8. Danby G et al (1962) Observation of high-energy neutrino reactions and the existence of two kinds of neutrinos. Phys Rev Lett 9:36–44. https://doi.org/10.1103/PhysRevLett.9.36
9. Kodama K et al (2001) (DONUT) Observation of tau neutrino interactions. Phys Lett B 504:218–224. https://doi.org/10.1016/S0370-2693(01)00307-0, hep-ex/0012035
10. Lesov A (2009) The weak force: from fermi to Feynman. arXiv:0911.0058
11. Garwin RL, Lederman LM, Weinrich M (1957) Observations of the failure of conservation of parity and charge conjugation in meson decays: the magnetic moment of the free muon. Phys Rev (U.S.) Superseded in part by Phys Rev A, Phys Rev B: Solid State, Phys Rev C, and Phys Rev D 105. https://doi.org/10.1103/PhysRev.105.1415
12. Wu CS et al (1957) Experimental test of parity conservation in beta decay. Phys Rev 105:1413–1415. https://doi.org/10.1103/PhysRev.105.1413
13. Feynman RP, Gell-Mann M (1958) Theory of the Fermi interaction. Phys Rev 109:193–198. https://doi.org/10.1103/PhysRev.109.193
14. Sudarshan ECG, Marshak RE (1958) Chirality Invariance and the universal Fermi interaction. Phys Rev 109:1860–1862. https://doi.org/10.1103/PhysRev.109.1860.2
15. Sakurai JJ (1958) Mass reversal and weak interactions. Il Nuovo Cimento (1955–1965) 7(5):649–660. https://doi.org/10.1007/BF02781569
16. Goldhaber M, Grodzins L, Sunyar AW (1958) Helicity of neutrinos. Phys Rev 109:1015–1017. https://doi.org/10.1103/PhysRev.109.1015
17. Atkinson RdE, Houtermans FG (1929) Zur Frage der Aufbaumöglichkeit der Elemente in Sternen. Zeitschrift für Physik 54(9):656–665. https://doi.org/10.1007/BF01341595
18. Bethe HA (1939) Energy production in stars. Phys Rev 55:434–456. https://doi.org/10.1103/PhysRev.55.434
19. Margaret Burbidge E, Burbidge GR, Fowler WA, Hoyle F (1957) Synthesis of the elements in stars. Rev Mod Phys 29:547–650. https://doi.org/10.1103/RevModPhys.29.547
20. Giunti C (2010) No effect of majorana phases in neutrino oscillations (January), pp. 1–6. arXiv:1001.0760v2
21. Berry C (2015) Blog entry December 2015. https://cplberry.com/2015/12/
22. Raymond D (1964) Solar neutrinos. II. Experimental. Phys Rev Lett 12:303–305. https://doi.org/10.1103/PhysRevLett.12.303
23. Bahcall John N (1964) Solar neutrinos. I. Theoretical. Phys Rev Lett 12:300–302. https://doi.org/10.1103/PhysRevLett.12.300
24. Bahcall JN (1989) Neutrino astrophysics. Cambridge University Press. https://books.google.co.uk/books?id=8GIP7uNMhlsC

25. Raymond D, Harmer Don S, Hoffman Kenneth C (1968) Search for neutrinos from the Sun. Phys Rev Lett 20:1205–1209. https://doi.org/10.1103/PhysRevLett.20.1205
26. Bahcall JN (1969) Neutrinos from the Sun. Sci Am 221(1):28–37. http://www.jstor.org/stable/24926407
27. Bahcall JN, Neutrino energy spectra. http://www.sns.ias.edu/~jnb/SNviewgraphs/SNspectrum/energyspectra.html
28. Fukuda Y et al (1996) Solar neutrino data covering solar cycle 22. Phys Rev Lett 77:1683–1686. https://doi.org/10.1103/PhysRevLett.77.1683
29. Hosaka J et al (2006) (Super-Kamiokande Collaboration) Solar neutrino measurements in Super-Kamiokande-I. Phys Rev D 73:112001. https://doi.org/10.1103/PhysRevD.73.112001
30. Lowe AJ (2009) Neutrino physics and the solar neutrino problem. arXiv:0907.3658
31. Cribier M et al (1999) Results of the whole GALLEX experiment. Nucl Phys B - Proc Suppl 70(1):284–291. Proceedings of the fifth international workshop on topics in astroparticle and underground physics. https://doi.org/10.1016/S0920-5632(98)00438-1
32. Abdurashitov JN et al (2002) (SAGE) Solar neutrino flux measurements by the Soviet-American Gallium Experiment (SAGE) for half the 22 year solar cycle. J Exp Theor Phys 95:181–193. https://doi.org/10.1134/1.1506424[astro-ph/0204245]
33. Bahcall JN, Solar neutrino view graphs. http://www.sns.ias.edu/~jnb/SNviewgraphs/snviewgraphs.html
34. Anchordoqui L, Paul T, Reucroft S, Swain J (2003) Ultrahigh energy cosmic rays. Int J Mod Phys A 18:2229–2366. https://doi.org/10.1142/S0217751X03013879[hep-ph/0206072]
35. Patrignani C et al (2016) (Particle Data Group) Review of particle physics. Chin Phys C40(10):100001. https://doi.org/10.1088/1674-1137/40/10/100001
36. Goldhaber M et al (1979) A proposal to test for baryon stability to a lifetime of 10**33-years
37. Becker-Szendy R et al (1992) Electron- and muon-neutrino content of the atmospheric flux. prd 46:3720–3724. https://doi.org/10.1103/PhysRevD.46.3720
38. Hirata KS et al (1992) Observation of a small atmospheric v/v_e ratio in Kamiokande. Phys Lett B 280:146–152. https://doi.org/10.1016/0370-2693(92)90788-6
39. Fukuda S et al (2003) The Super-Kamiokande detector. Nucl Instrum Methods Phys Res Sect A: Accel Spectrom Detect Assoc Equip 501(2):418–462. https://doi.org/10.1016/S0168-9002(03)00425-X
40. Fukuda Y et al (1998) Measurement of a small atmospheric ν_μ/ν_e ratio. Phys Lett B 433(1):9–18. https://doi.org/10.1016/S0370-2693(98)00476-6
41. Shiozawa M (1999) Reconstruction algorithms in the Super-Kamiokande large water Cherenkov detector. Nucl Instrum Methods Phys Res Sect A: Accel Spectrom Detect Assoc Equip 433(1):240–246. https://doi.org/10.1016/S0168-9002(99)00359-9
42. Kajita T, Kearns E, Shiozawa M (2016) Establishing atmospheric neutrino oscillations with Super-Kamiokande. Nucl Phys B 908(Supplement C):14–29. Neutrino oscillations: celebrating the Nobel Prize in Physics (2015). https://doi.org/10.1016/j.nuclphysb.2016.04.017, http://www.sciencedirect.com/science/article/pii/S0550321316300554
43. Pontecorvo B (1957) Mesonium and anti-mesonium. Sov Phys JETP 6:429. [Zh Eksp Teor Fiz 33:549(1957)]
44. Ziro M, Masami N, Shoichi S (1962) Remarks on the unified model of elementary particles. Prog Theor Phys 28(5):870–880. https://doi.org/10.1143/PTP.28.870
45. Pontecorvo B (1968) Neutrino experiments and the problem of conservation of leptonic charge. Sov Phys JETP 26:984–988. [Zh Eksp Teor Fiz 53:1717 (1967)]
46. Eliezer S, Swift AR (1976) Experimental consequences of electron neutrino-muon-neutrino mixing in neutrino beams. Nucl Phys B 105:45–51. https://doi.org/10.1016/0550-3213(76)90059-6
47. Fritzsch H, Minkowski P (1976) Vector-like weak currents, massive neutrinos, and neutrino beam oscillations. Phys Lett 62B:72–76. https://doi.org/10.1016/0370-2693(76)90051-4
48. Bilenky SM, Pontecorvo B (1976) Again on neutrino oscillations. Lett Nuovo Cim 17:569. https://doi.org/10.1007/BF02746567

49. Wolfenstein L (1978) Neutrino oscillations in matter. Phys Rev D 17:2369–2374. https://doi.org/10.1103/PhysRevD.17.2369
50. Mikheev SP, Smirnov AYu (1985) Resonance amplification of oscillations in matter and spectroscopy of solar neutrinos. Sov J Nucl Phys 42:913–917. [Yad Fiz 42:1441 (1985)]
51. Mikheev SP, Smirnov AY (1986) Resonant amplification of neutrino oscillations in matter and solar neutrino spectroscopy. Nuovo Cim C 9:17–26. https://doi.org/10.1007/BF02508049
52. Bilenky S, Giunti C, Grimus W (1999) Phenomenology of neutrino oscillations. Prog Part Nucl Phys 43:1–86. https://doi.org/10.1016/S0146-6410(99)00092-7, hep-ph/9812360
53. Smirnov AYu (2005) The MSW effect and matter effects in neutrino oscillations. Phys Scripta T121:57–64. https://doi.org/10.1088/0031-8949/2005/T121/008, hep-ph/0412391
54. Mikheev SP, Smirnov AYu (1986) Neutrino oscillations in a variable density medium and neutrino bursts due to the gravitational collapse of stars. Sov Phys JETP 64:4–7. arXiv:0706.0454, [Zh Eksp Teor Fiz 91:7 (1986)]
55. Fukuda Y et al (1998) (Super-Kamiokande) Evidence for oscillation of atmospheric neutrinos. Phys Rev Lett 81:1562–1567. https://doi.org/10.1103/PhysRevLett.81.1562[hep-ex/9807003]
56. Barger Vernon D et al (1999) Neutrino decay and atmospheric neutrinos. Phys Lett B 462:109–114. https://doi.org/10.1016/S0370-2693(99)00887-4[hep-ph/9907421]
57. Yuval G, Worah MP (1998) Atmospheric muon-neutrino deficit from decoherence. Phys Lett B [hep-ph/9807511]. Submitted to
58. Bellerive A et al, The sudbury neutrino observatory, pp 1–25, arXiv:1602.02469v2
59. Boger J et al (2000) The sudbury neutrino observatory, vol 449, pp 172–207
60. Amsbaugh John F et al (2007) An array of low-background He-3 proportional counters for the sudbury neutrino observatory. Nucl Instrum Meth A579:1054–1080. https://doi.org/10.1016/j.nima.2007.05.321, arXiv:0705.3665
61. Ahmad QR et al (2002) (SNO) Direct evidence for neutrino flavor transformation from neutral current interactions in the sudbury neutrino observatory. Phys Rev Lett 89:011301. https://doi.org/10.1103/PhysRevLett.89.011301[nucl-ex/0204008]
62. Smirnov AYu (2016) Solar neutrinos: oscillations or no-oscillations? arXiv:1609.02386
63. Araki T et al (2005) (KamLAND) Measurement of neutrino oscillation with KamLAND: evidence of spectral distortion. Phys Rev Lett 94:081801. https://doi.org/10.1103/PhysRevLett.94.081801[hep-ex/0406035]
64. Abe S et al (2008) (KamLAND) Precision measurement of neutrino oscillation parameters with KamLAND. Phys Rev Lett 100:221803. https://doi.org/10.1103/PhysRevLett.100.221803, arXiv:0801.4589
65. de Salas PF et al (2017) Status of neutrino oscillations 2017. arXiv:1708.01186
66. Ahn JK et al (2012) (RENO) Observation of reactor electron antineutrino disappearance in the RENO experiment. Phys Rev Lett 108:191802. https://doi.org/10.1103/PhysRevLett.108.191802, arXiv:1204.0626
67. Ahn JK et al (2010) (RENO) RENO: an experiment for neutrino oscillation parameter θ_{13} using reactor neutrinos at Yonggwang. arXiv:1003.1391
68. Abe Y et al (2014) (Double Chooz) Improved measurements of the neutrino mixing angle θ_{13} with the Double Chooz detector. JHEP 10:086. https://doi.org/10.1007/JHEP02(2015)074, https://doi.org/10.1007/JHEP10(2014)086, arXiv:1406.7763
69. Adamson P et al (2014) (MINOS) Combined analysis of $\nu_{\mu''}$ disappearance and $\nu_\mu \rightarrow \nu_e$ appearance in MINOS using accelerator and atmospheric neutrinos. Phys Rev Lett 112:191801. https://doi.org/10.1103/PhysRevLett.112.191801, arXiv:1403.0867
70. Abe K et al (2017) (T2K Collaboration) Combined analysis of neutrino and antineutrino oscillations at T2K. Phys Rev Lett 118:151801. https://doi.org/10.1103/PhysRevLett.118.151801
71. Adamson P et al (2017) (NOvA) Constraints on oscillation parameters from ν_e appearance and ν_μ disappearance in NOvA. Phys Rev Lett 118(23):231801. https://doi.org/10.1103/PhysRevLett.118.231801, arXiv:1703.03328

72. Abe K et al (2013) (Super-Kamiokande) Evidence for the appearance of atmospheric Tau neutrinos in Super-Kamiokande. Phys Rev Lett 110(18):181802. https://doi.org/10.1103/PhysRevLett.110.181802, arXiv:1206.0328
73. Wendell R et al (2010) (Super-Kamiokande) Atmospheric neutrino oscillation analysis with sub-leading effects in Super-Kamiokande I, II, and III. Phys Rev D81:092004. https://doi.org/10.1103/PhysRevD.81.092004, arXiv:1002.3471
74. Aartsen MG et al (2017) (IceCube) Measurement of atmospheric neutrino oscillations at 6–56 GeV with icecube deepcore. arXiv:1707.07081
75. Abbasi R et al (2012) (IceCube) The design and performance of icecube deepcore. Astropart Phys 35:615–624. https://doi.org/10.1016/j.astropartphys.2012.01.004, arXiv:1109.6096
76. Adrian-Martinez S et al (2013) (ANTARES) Measurement of the atmospheric ν_μ energy spectrum from 100 GeV to 200 TeV with the ANTARES telescope. Eur Phys J C73(10):2606. https://doi.org/10.1140/epjc/s10052-013-2606-4, arXiv:1308.1599
77. Aharmim B et al (2013) (SNO) Combined analysis of all three phases of solar neutrino data from the sudbury neutrino observatory. Phys Rev C88:025501. https://doi.org/10.1103/PhysRevC.88.025501, arXiv:1109.0763
78. Abe K et al (2016) (Super-Kamiokande) Solar neutrino measurements in Super-Kamiokande-IV. Phys Rev D94(5):052010. https://doi.org/10.1103/PhysRevD.94.052010, arXiv:1606.07538
79. Bellini G et al (2012) (Borexino Collaboration) First evidence of *pep* solar neutrinos by direct detection in Borexino. Phys Rev Lett 108:051302. https://doi.org/10.1103/PhysRevLett.108.051302
80. Bellini G et al (2011) (Borexino Collaboration) Precision measurement of the ^{7}Be solar neutrino interaction rate in Borexino. Phys Rev Lett 107:141302. https://doi.org/10.1103/PhysRevLett.107.141302
81. Bellini G et al (2010) (Borexino Collaboration) Measurement of the solar ^{8}B neutrino rate with a liquid scintillator target and 3 MeV energy threshold in the Borexino detector. Phys Rev D 82:033006. https://doi.org/10.1103/PhysRevD.82.033006
82. Smirnov OYu et al (2016) (Borexino) Measurement of neutrino flux from the primary proton-proton fusion process in the Sun with Borexino detector. Phys Part Nucl 47(6):995–1002. https://doi.org/10.1134/S106377961606023X, arXiv:1507.02432
83. Patterson RB (2015) Prospects for measurement of the neutrino mass hierarchy. Ann Rev Nucl Part Sci 65:177–192. https://doi.org/10.1146/annurev-nucl-102014-021916, arXiv:1506.07917
84. Martn-Albo J (DUNE) (2017) Sensitivity of DUNE to long-baseline neutrino oscillation physics. In: 2017 European physical society conference on high energy physics (EPS-HEP 2017) Venice, Italy. https://inspirehep.net/record/1632445/files/arXiv:1710.08964.pdf, arXiv:1710.08964. Accessed 5–12 July 2017
85. Strait J et al (2016) (DUNE) Long-baseline neutrino facility (LBNF) and deep underground neutrino experiment (DUNE). arXiv:1601.05823
86. Kim S-B (2015) (RENO) New results from RENO and prospects with RENO-50. Nucl Part Phys Proc 265-266:93–98. https://doi.org/10.1016/j.nuclphysbps.2015.06.024, arXiv:1412.2199
87. Fengpeng A et al (2016) (JUNO) Neutrino physics with JUNO. J Phys G43(3):030401. https://doi.org/10.1088/0954-3899/43/3/030401, arXiv:1507.05613
88. Hongxin W et al (2017) Mass hierarchy sensitivity of medium baseline reactor neutrino experiments with multiple detectors. Nucl Phys B 918:245–256. https://doi.org/10.1016/j.nuclphysb.2017.03.002, arXiv:1602.04442
89. Liang Z, Yifang W, Jun C, Liangjian W (2009) Experimental requirements to determine the neutrino mass hierarchy using reactor neutrinos. Phys Rev D 79:073007. https://doi.org/10.1103/PhysRevD.79.073007
90. Yu-Feng L, Jun C, Yifang W, Liang Z (2013) Unambiguous determination of the neutrino mass hierarchy using reactor neutrinos. Phys Rev D 88:013008. https://doi.org/10.1103/PhysRevD.88.013008

91. Yokoyama M (2017) (Hyper-Kamiokande Proto) The hyper-Kamiokande experiment. In: Proceedings, prospects in neutrino physics (NuPhys2016), London, UK. arXiv:1705.00306. Accessed 12–14 Dec 2016

92. Walter W (2013) Neutrino mass hierarchy determination with IceCube-PINGU. Phys Rev D88(1):013013. https://doi.org/10.1103/PhysRevD.88.013013, arXiv:1305.5539

93. Clark K (2016) PINGU and the neutrino mass hierarchy. Nucl Part Phys Proc 273–275(Supplement C):1870–1875. 37th International conference on high energy physics (ICHEP). https://doi.org/10.1016/j.nuclphysbps.2015.09.302, http://www.sciencedirect.com/science/article/pii/S2405601415007919

94. Capozzi F, Lisi E, Marrone A (2017) Probing the neutrino mass ordering with KM3NeT-ORCA: Analysis and perspectives. arXiv:1708.03022

95. JUNO Collaboration/JGU-Mainz, Neutrino mass hierarchy in JUNO. https://next.ific.uv.es/next/experiment/physics.html

96. Bron S (2015) Hyper-Kamiokande: towards a measurement of CP violation in lepton sector. In: Proceedings, prospects in neutrino physics (NuPhys2015), London, UK. arXiv:1605.00884. Accessed 16–18 Dec 2015

97. Drexlin G, Hannen V, Mertens S, Weinheimer C (2013) Current direct neutrino mass experiments. Adv High Energy Phys 2013:293986. https://doi.org/10.1155/2013/293986, arXiv:1307.0101

98. Formaggio JA (2014) Direct neutrino mass measurements after PLANCK. Phys Dark Univ 4(Supplement C):75–80. https://doi.org/10.1016/j.dark.2014.10.004, dARK TAUP2013, http://www.sciencedirect.com/science/article/pii/S2212686414000314

99. Hamish Robertson RG (2013) (KATRIN) KATRIN: an experiment to determine the neutrino mass from the beta decay of tritium. In: Proceedings, 2013 community summer study on the future of U.S. Particle physics: snowmass on the mississippi (CSS2013), Minneapolis, MN, USA. arXiv:1307.5486. Accessed July 29–Aug 6 2013

100. Esfahani AA et al (2017) (Project 8) Determining the neutrino mass with cyclotron radiation emission spectroscopyProject 8. J Phys G44(5):054004. https://doi.org/10.1088/1361-6471/aa5b4f, arXiv:1703.02037

101. Blaum K et al (2013) The electron capture ^{163}Ho experiment ECHo. In: The future of neutrino mass measurements: terrestrial, astrophysical, and cosmological measurements in the next decade (NUMASS2013) Milano, Italy. arXiv:1306.2655, Accessed 4–7 Feb 2013

102. Couchot F et al (2017) Cosmological constraints on the neutrino mass including systematic uncertainties. Astron Astrophys https://doi.org/10.1051/0004-6361/201730927, arXiv:1703.10829 [Astron Astrophys 606:A104 (2017)]

103. Coleman S (2011) Notes from Sidney Coleman's physics 253a: quantum field theory. arXiv:1110.5013

104. Palash B (2011) Pal, Dirac, Majorana and Weyl Fermions. Am J Phys 79:485–498. https://doi.org/10.1119/1.3549729, arXiv:1006.1718

105. Thomson M (2013) Modern particle physics. Cambridge University Press, New York. http://www-spires.fnal.gov/spires/find/books/www?cl=QC793.2.T46::2013

106. Laurent C, Marco D, Mikhail S (2012) Matter and antimatter in the universe. New J Phys 14:095012. https://doi.org/10.1088/1367-2630/14/9/095012, arXiv:1204.4186

107. Ade PAR et al (Planck) (2016) Planck 2015 results. XIII. Cosmological parameters. Astron Astrophys 594:A13. https://doi.org/10.1051/0004-6361/201525830, arXiv:1502.01589

108. Sakharov AD (1967) Violation of CP invariance, c Asymmetry, and baryon asymmetry of the universe. Pisma Zh Eksp Teor Fiz 5:32–35. https://doi.org/10.1070/PU1991v034n05ABEH002497, [Usp Fiz Nauk 161:61 (1991)]

109. Fukugita M, Yanagida T (1986) Baryogenesis without grand unification. Phys Lett B 174:45–47. https://doi.org/10.1016/0370-2693(86)91126-3

110. Kuzmin VA, Rubakov VA, Shaposhnikov ME (1985) On anomalous electroweak baryon-number non-conservation in the early universe. Phys Lett B 155(1):36–42. https://doi.org/10.1016/0370-2693(85)91028-7

111. Petcov ST, The nature of massive neutrinos, vol 1 pp 1–35, arXiv:1303.5819v1

112. Langacker P, Petcov ST, Steigman G, Toshev S (1987) On the Mikheev-Smirnov-Wolfenstein (MSW) mechanism of amplification of neutrino oscillations in matter. Nucl Phys B 282:589–609. https://doi.org/10.1016/0550-3213(87)90699-7

113. Dell'Oro S, Marcocci S, Viel M, Vissani F (2016) Neutrinoless double beta decay: 2015 review. Adv High Energy Phys 2016:2162659. https://doi.org/10.1155/2016/2162659, arXiv:1601.07512

114. Goeppert-Mayer M (1935) Double beta-disintegration. Phys Rev 48:512–516. https://doi.org/10.1103/PhysRev.48.512

115. Tretyak VI, Zdesenko YG (2002) Tables of double beta decay dataan update. At Data Nucl Data Tables 80(1): 83–116

116. Turkevich AL, Economou T, Cowan GA (1992) Double-beta decay of U 238, vol 67 pp 3211–3214

117. Elliott SR, Hahn AA, Moe MK (1987) Direct evidence for two-neutrino double-beta decay in ^{82}Se. Phys Rev Lett 59:2020–2023. https://doi.org/10.1103/PhysRevLett.59.2020

118. Alduino C et al (2017) Measurement of the two-neutrino double-beta decay half-life of 130 Te with the CUORE-0 experiment. Eur Phys J C. https://doi.org/10.1140/epjc/s10052-016-4498-6

119. Furry WH (1939) On transition probabilities in double beta-disintegration. Phys Rev 56:1184–1193. https://doi.org/10.1103/PhysRev.56.1184

120. Racah G, Nakagawa K (1982) On the symmetry between particles and antiparticles. Soryushiron Kenkyu 65(1):34–41. http://ci.nii.ac.jp/naid/110006468258/en/

121. Schechter J, Valle JWF (1982) Neutrinoless double-β decay in $SU(2) \times U(1)$ theories. Phys Rev D 25:2951–2954. https://doi.org/10.1103/PhysRevD.25.2951

122. Takasugi E (1984) Can the neutrinoless double beta decay take place in the case of Dirac neutrinos? Phys Lett B 149(4–5):372–376

123. Heinrich P, Werner R (2015) Neutrinoless double beta decay. New J Phys 17(11):115010. https://doi.org/10.1088/1367-2630/17/11/115010, arXiv:1507.00170

124. Engel Jonathan, Menndez Javier (2017) Status and future of nuclear matrix elements for neutrinoless double-beta decay: a review. Rep Prog Phys 80(4):046301. https://doi.org/10.1088/1361-6633/aa5bc5, arXiv:1610.06548

125. Benato Giovanni (2015) Effective Majorana Mass and Neutrinoless Double Beta Decay. Eur. Phys. J. C75(11):563. https://doi.org/10.1140/epjc/s10052-015-3802-1, 1510.01089

126. Gando A et al (2016) (KamLAND-Zen) Search for majorana neutrinos near the inverted mass hierarchy region with KamLAND-Zen. Phys Rev Lett 117(8):082503. https://doi.org/10.1103/PhysRevLett.117.109903, https://doi.org/10.1103/PhysRevLett.117.082503, [Addendum: Phys Rev Lett 117(10):109903 (2016)]

127. Tretyak VI, Zdesenko YuG (1995) Tables of double beta decay data. At Data Nucl Data Tables 61(1):43–90. https://doi.org/10.1016/S0092-640X(95)90011-X

128. Bellini G (2016) The impact of Borexino on the solar and neutrino physics. Nucl Phys B 908(Supplement C):pp 178–198. Neutrino oscillations: celebrating the Nobel Prize in physics 2015. https://doi.org/10.1016/j.nuclphysb.2016.04.011, http://www.sciencedirect.com/science/article/pii/S0550321316300499

129. Elliott SR, Vogel P (2002) Double beta decay. Ann Rev Nucl Part Sci 52:115–151

130. Biller SD (2013) Probing Majorana neutrinos in the regime of the normal mass hierarchy. Phys Rev D87(7):071301. https://doi.org/10.1103/PhysRevD.87.071301, arXiv:1306.5654

131. Barabash AS (2017) Brief review of double beta decay experiments. arXiv:1702.06340

132. Alduino C et al (CUORE) First results from CUORE: a search for lepton number violation via $0\nu\beta\beta$ decay of ^{130}Te. arXiv:1710.07988

133. The GERDA Collaboration, "Background-free search for neutrinoless double-decay of 76Ge with GERDA", Nature vol. 544 pp. 47 EP – (2017), URL http://dx.doi.org/10.1038/nature21717

134. Agostini M et al (2018) (GERDA Collaboration) Improved limit on neutrinoless double-β decay of ^{76}Ge from GERDA phase II. Phys Rev Lett 120:132503. https://doi.org/10.1103/PhysRevLett.120.132503

135. Albert JB et al (2014) (EXO-200) Search for Majorana neutrinos with the first two years of EXO-200 data. Nature 510:229–234. https://doi.org/10.1038/nature13432, arXiv:1402.6956

136. Arnold R et al (2005) (NEMO) First results of the search of neutrinoless double beta decay with the NEMO 3 detector. Phys Rev Lett 95:182302. https://doi.org/10.1103/PhysRevLett.95.182302[hep-ex/0507083]

137. NEMO Collaboration et al (1996) Double-decay of116Cd, vol 72 pp 239–247

138. Bongrand M, Mesure des processus de double desintegration beta du 130 Te dans l'experience NEMO 3 R&D du projet SuperNEMO : etude d'un detecteur BiPo, Ph.D. thesis

139. Argyriades J et al (2009) Measurement of the double-beta decay half-life of 150Nd and search for neutrinoless decay modes with the NEMO-3 detector, vol 80

140. Argyriades J et al (2010) Measurement of the two neutrino double beta decay half-life of Zr-96 with the NEMO-3 detector. Nucl Phys A 847(3):168–179. https://doi.org/10.1016/j.nuclphysa.2010.07.009, http://www.sciencedirect.com/science/article/pii/S0375947410006238

141. Arnold R et al (2016) (NEMO-3) Measurement of the double-beta decay half-life and search for the neutrinoless double-beta decay of ^{48}Ca with the NEMO-3 detector. Phys Rev D93(11):112008. https://doi.org/10.1103/PhysRevD.93.112008, arXiv:1604.01710

142. Arnold R et al (2010) Probing new physics models of neutrinoless double beta decay with SuperNEMO. Eur Phys J C 70(4):927–943. https://doi.org/10.1140/epjc/s10052-010-1481-5

143. Andringa S et al (2016) Current status and future prospects of the SNO + experiment. arXiv:1508.05759v3 [physics.ins-det]. Accessed 28 Jan 2016

144. Wright A (2017) Te Dev R&D pre-summary. SNO+-docDB 4545-v1

Chapter 2
The SNO+ Experiment

The SNO+ experiment will retrofit the SNO detector [1] by replacing its heavy water target with a liquid scintillator one. The SNO detector has several features that are ideal for new investigations. First, its location 2 km underground in the Creighton Mine, Sudbury, Canada ensures a muon flux of just $0.27 \, \mu/m^2/day$ [2]. Second, it still has the highest photo-cathode coverage of any kTonne scale water Cherenkov or liquid scintillator detector [1, 3–5]. Finally, the experiment now sits within SNOLAB, a multi-purpose laboratory providing excellent infrastructure for providing the radio-purity that is critical to many rare process searches.

Transitioning from a water Cherenkov detector to a liquid scintillator one provides several advantages. First, organic scintillators are non-polar and therefore purifiable to 1000 times smaller concentrations of radio-impurities than water [1, 6]. Second, organic scintillators produce more photons per MeV of deposited energy than Cherenkov emission in water by as much as $50\times$. This makes it possible to lower the trigger threshold to sub-MeV energies, and to achieve energy resolutions of a few $\% \cdot \sqrt{(E/MeV)}$ [3, 5].

To facilitate the transition, the detector has been fitted with 'hold down' ropes to counter the buoyancy of the scintillator relative to its water bath shield and the DAQ system has been upgraded for the increased light yield and higher event rates, as well as to accept the broader time profile expected from scintillation light [7].

The experiment will progress through 3 phases, filling the target volume first with water, then pure liquid scintillator, then, finally, tellurium loaded liquid scintillator [7]. The primary goal of the SNO+ experiment and the primary concern of this work is the search for $0\nu\beta\beta$ in ^{130}Te. Here, the significance of the initial two phases is their potential to provide constraints for the final tellurium phase but they have a physics reach of their own that is better described elsewhere [7].

This chapter describes the SNO+ hardware as an extremely sensitive photon counter, reviewing the creation and propagation of optical photons inside it. A brief overview of the expected calibration campaign explains how unknowns in the detector model are constrained and, finally, the detector simulation used to apply that model to a search for $0\nu\beta\beta$ is discussed.

© Springer Nature Switzerland AG 2019
J. Dunger, *Event Classification in Liquid Scintillator Using PMT Hit Patterns*,
Springer Theses, https://doi.org/10.1007/978-3-030-31616-7_2

2.1 Detector

Optical photons created in the detector centre propagate outwards until they reach
photon detectors which record their arrival times. The SNO+ detector is organised
as a series of near concentric spheres around a central target, to maximise photon
detection probability. Figure 2.1 shows a cross section; working inwards it shows
[1, 7]:

- **Cavity**: This 22×34 m cylindrical void houses 7 kTonne of ultra-pure water
  (UPW). The water shields the detector from radiation originating in the surrounding rock and provides buoyancy to support it (external water onwards).
- **Photomultiplier Tubes (PMTs) and PMT support structure (PSUP)**: The PSUP
  is a geodesic sphere of radius 8.8 m built from triangular panels, each supporting
  10 s of PMTs. The PSUP itself if supported by cables running to the cavity walls
  and ceiling.
- **Hold up and hold down ropes**: These exist to offset differences in buoyancy
  between the target and the surrounding water shield. The hold-up ropes existed to

Fig. 2.1 A cut through of
the SNO+ detector showing
the PMT array, acrylic
vessel, hold up/down ropes,
the surrounding rock cavity
and the deck above [8]

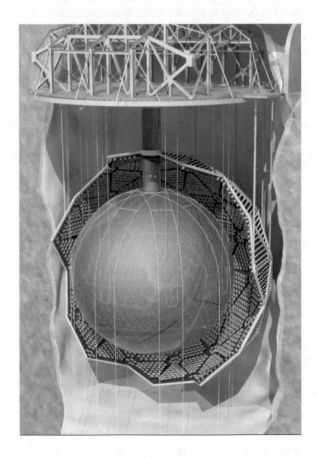

support a heavy water target in the SNO experiment. The hold-down ropes prevent
the new, positively buoyant, liquid scintillator target from floating upwards.

- **Acrylic vessel (AV)**: The AV is an optically polished sphere, 5.5 cm thick and 12 m
  across, which houses the water/liquid scintillator targets. It is filled via a 6.8 m long
  cylindrical neck at its north pole.
- **Cover gas**: The SNO cover gas system was upgraded to meet more stringent
  requirements for liquid scintillator. The detector is now kept isolated from the
  laboratory air by radon tight buffer bags filled with high purity nitrogen gas. The
  system sits between the upper surface of the external water and the laboratory
  above. The system is designed to reduce radon contamination by a factor of 10^5 [7].

2.2 PMTs

The basic detection unit of the SNO+ detector is the Hamamatsu 8″ R1408 PMT
shown in Fig. 2.2. A PMT uses the photo-electric effect and charge multiplication to
perform single photon counting. The main structure is glass containing a vacuum [9].
The inner surface of this glass is coated with a thin caesium bialkali film [9] called
the photo-cathode. The glass is held at ground potential, whereas the anode at the
base is held at 1700–2100 V, creating a strong field inside the PMT. The dynode stack
consists of 7 metal plates, coated with a secondary emissive material and horizontally
stacked into a 'venetian blind'. Together they form a potential divider circuit, which
steps up the voltage from ground to the anode potential.

A photon incident on the photo-cathode can create a photo-electron (p.e.). Electric
fields within the main volume accelerate any created p.e. towards the central dynode

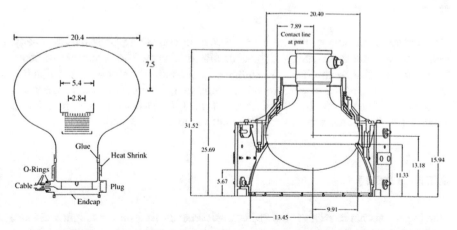

Fig. 2.2 Left: Cross-section of the R1408 PMT. The seven dynodes are shown as solid horizontal
lines, the anode is the 'U' at the bottom and the focussing grid is the dashed line above the dynodes.
Right: Cross-section of the PMT with attached concentrator. Measures in cm [1]

Fig. 2.3 Combined
collection and quantum
efficiency as a function of
wavelength for the
Hamamatsu R1408 PMT.
Each line shows an
individual R1408 PMT [10]

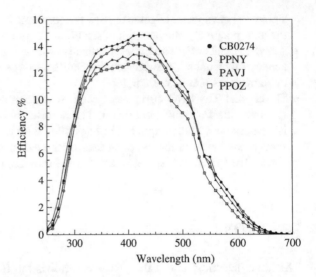

and focus them onto the stack. Potential differences accelerate electrons through the
stack and, at each stage, the total charge is multiplied by secondary electrons created
in collisions with the plates. This magnifies the current into a measurable signal that
is fed out to an external 'base' circuit at the anode.

Originally installed for the SNO experiment, these PMTs are now far from the
state of the art. What follows is a non-exhaustive review of the R1408 features with
significance for the studies to follow.

2.2.1 Efficiency

Two factors contribute to the PMTs efficiency in converting photons into charge
pulses: the quantum efficiency and the collection efficiency. The first is the efficiency
in creating p.e. that escape the photo-cathode, the second is the efficiency in collecting
them. i.e. what fraction of p.e. are successfully magnified down the dynode stack.
Figure 2.3 shows the combined efficiency of 4 R1408 PMTs [10] as a function of
wavelength. On average, the peak efficiency is around 13.5% at 440 nm.

2.2.2 Timing

One important characteristic is the PMT transit time: the time between light incidence
on the photo-cathode and the time the resulting current pulse reaches its maximum.
The absolute value of the transit time determines an offset on all PMT hits and
is therefore unimportant here. However, the jitter on this transit time, commonly

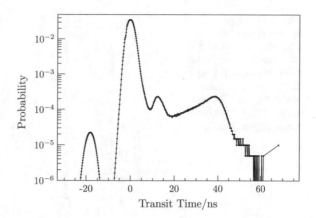

Fig. 2.4 The transit time response of a typical Hamamatsu R1408 PMT, taken from the RAT simulation, originally measured by P. Skensved

referred to as the transit time spread (TTS), is more important because it leads to ambiguities: is an early charge pulse the result of an early photon or a shorter than average transit time?

Figure 2.4 shows the transit time distribution for the R1408. The dominant 'prompt' peak has a FWHM of 3.7 ns. The sub-dominant early peak is 'pre-pulsing': cases where photons have passed through the photo-cathode but created a p.e. directly on the top of the dynode stack. Later 'late-pulsing' peaks arise when p.e. elastically scatter on the first dynode back towards the photo-cathode, only creating secondary electrons when they arrive at the top of the stack a second time.

2.2.3 Charge

Even with no incident photons, PMT produce current at their base. In SNO+, real p.e. are distinguished from this 'dark current' using the total charge deposited at the base. Figure 2.5 shows the single p.e. and dark current spectra for the R1408. A threshold is set at roughly 1/4 of the average charge deposited by a single electron, determined for each PMT separately. When this threshold is crossed the PMT is considered 'hit'. In this configuration, around 25% of real p.e. are lost, incorrectly labelled as noise, and the PMTs cross threshold on noise at a rate of 500 Hz [1].

In principle, the total charge deposited can also used to estimate the number of p.e. arriving in a single PMT. Figure 2.5 shows the distribution of the total charge (in ADC counts) read out at the anode, for 1–4 p.e. created in the PMT during a 400 ns period. Unfortunately, at low numbers of p.e., it is near impossible to discriminate to better than ±2 p.e. from charge alone.

Fig. 2.5 Charge
performance of the
Hamamatsu R1408 PMT.
Top: the single p.e. spectrum
with dark noise [1], typical
thresholds are 9 counts
above pedestal. Bottom:
analytically calculated multi
p.e. charge responses [11]

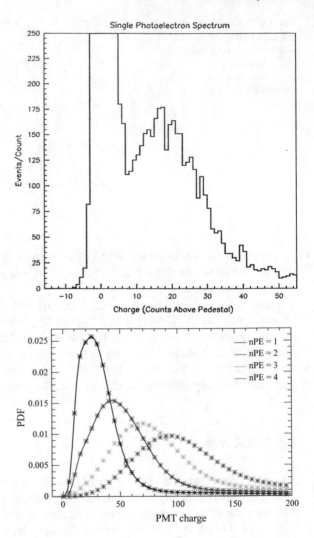

2.2.4   Arrangement

Photons can only be detected if they hit a PMT, so the 9700 PMTs are arranged
on the PSUP facing inwards to maximise photo-cathode coverage of the 4π solid
angle, as viewed from the detector centre [1]. In addition, each PMT is fitted with a
'concentrator', a bucket built from reflective petals and placed in front of the PMT
face (see Fig. 2.2). These concentrators increase the photo-cathode coverage from
31 to 59% by area [1]. Of course, the reflectors are not 100% reflective, and so the
effective coverage is around 54% [7]. Moreover, the concentrators aged during SNO,
reducing the effective coverage to 49%, and around 400 PMTs failed, leaving around
9300 active PMTs [7].

There are also 4 PMTs surrounding the neck to help veto events occurring inside it, and 91 outward looking PMTs to veto events from high energy particles, mostly muons, passing through the external water.

2.3 Data Acquisition System (DAQ)

The DAQ system is responsible for reading out the time and charge of PMT hits and grouping them into events. Most simply, this is achieved by constantly buffering the data associated with each PMT hit, until a global trigger arrives to collect it or a reset time elapses and it is discarded. Global triggers (GT) are issued on the coincidence of several PMT hits, determined by the analogue sum of trigger signals issued by each PMT that crosses threshold. The next sections describe the most important parts of this system's implementation.[1]

2.3.1 PMT Interface

Each PMT is connected by a single cable that both supplies the high voltage and carries charge signals from the PMT: this is an electronics 'channel'. The 9728 channels are divided into groups of 512 channels, each managed by one of 18 identical electronics crates.

A single crate contains 16 PMT interface cards (PMTIC), with 4 attached paddle cards. Every channel meets the crate at one of these paddle cards, giving 8 channels per paddle, 32 per PMTIC.

The role the PMTIC is two fold. First, it controls the high voltage supply to each PMT via 4 relays that control the supply to 8 PMTs at a time. Second, it collects signals from the 32 PMTs and forwards them on to the readout system.

2.3.2 Readout

Inside the crate, each PMTIC has an associated front end card (FEC) which, mirroring the PMTIC paddles, connects to 4 daughter boards (DB), each responsible for 8 channels. Signals passed from the PMTIC arrive at one of these 4 DBs.

The DB are responsible for the discrimination and integration of the PMT signals. Each is fitted with two 4 channel discriminators, two 8 channel integrators and 8 1 channel CMOS chips. If an input signal crosses a channel's threshold, a hit is registered and a time-to-analogue-converter (TAC) inside the channel's CMOS chip starts. At the same time, the input signal is passed into two of the integration channels,

---

[1]A more complete description of the SNO DAQ may be found in [1]. The SNO+ upgrades are described in [7].

one high gain and one low gain, which integrate over short and long time intervals (60 and 390 ns, respectively).

If a global trigger arrives, the TAC stops and, after the integration time is complete, the charge and time (relative to trigger) is saved to analogue memory cells in the CMOS chip and a 'CMOS data available' flag is set. A hex ID number for the GT (GTID) and status flags for the hit are saved in digital memory. If no GT arrives within 410 ns, the data are not saved and the TAC resets automatically.

The FEC card itself contains a sequencer, analogue to digital converters (ADCs), and FIFO memory. The FEC sequencer continuously polls all of the 'data available' flags. If a flag is set, the sequencer pulls the hit data from memory and sends it to an ADC for digitisation. The output is a PMT bundle: three 8 bit words containing, among other things, the GTID, addresses for the PMT, the digitised charge integrals and the hit time. This bundle is pushed into the FIFO and a 'FEC data available' flag is set.

Each crate also contains a single XL3 card, connected to each of the FECs via a backplane. One major responsibility of the XL3 card is to monitor the FEC data flags, retrieve the PMT bundles from the FEC and bus them over Ethernet to a dedicated server.

2.3.3 Trigger

When a channel crosses threshold, its DB injects analogue signals into several trigger sums. The sum is performed at the crate level by a dedicated crate trigger card (CTC). Each CTC, in turn, passes its total on to 7 MTC/A+ boards where they are combined into a detector wide sum. Each MTC/A+ is equipped with three amplifiers and three corresponding discriminators with adjustable thresholds. If one of these discriminators fires, a signal is passed on to the MTC/D. The MTC/D compares this trigger type against a programmable mask. If the type of the GT is masked in, the MTC/D issues a GT latched to its 50 MHz clock. The trigger gate is 400 ns long, collecting all of the hits registered in the 400 ns before its arrival. After a GT is issued, there is a 420 ns lockout during which no further GTs may be issued.

The main workhorse physics trigger is N100. For this trigger, each PMT hit issues a square pulse of length 100 ns and equal height. Thresholding on the sum of these pulses detects coincidences of a set number of hits within a 100 ns window.

The MTC/D can force individual channel discriminators to fire (PED) or manually issue global triggers (PULGT) and synchronously/asynchronously accept external triggers from calibration sources.

2.4 The Target

The SNO+ collaboration has selected Linear Alkyl Benzene (LAB) as its primary scintillator. It has excellent light levels [12], good $\alpha-\beta$ discrimination [13], it is compatible with acrylic [7] and it is extremely cheap. At the time of writing, a purpose built scintillator plant is being commissioned to purify the required 782 tonnes of LAB underground [7, 14]. The plant uses the same techniques as the Borexino plant [6, 14] and similarly targets purification levels of 10^{-17} g of uranium/thorium chain contaminants per g of LAB. At the end of the water phase, scintillator will be fed in to the top of the AV, gradually replacing water removed using pipes at its base.

One of the biggest experimental challenges for SNO+ is loading the tellurium into the liquid scintillator. This is difficult for exactly the same reason that the scintillator is easy to purify: its non-polarity. Furthermore, the loading must be done in a way that is compatible with acrylic and other detector components, that is safe for underground laboratories, and which preserves the optics and radio-purity of the scintillator.

This will be achieved by first manufacturing an organometallic complex of tellurium with 1,2 butanediol (Diol and TeDiol onwards). TeDiol is soluble in LAB, highly transparent, compatible with acrylic and it can be easily purified with distillation [15, 16].

The tellurium is procured as telluric acid (TeA). After purchase, the TeA will be shipped underground where it will be purified with acid-base extraction using nitric acid and synthesised into TeDiol.

Phase I of SNO+ will load 0.5% tellurium by mass into the scintillator volume using 1.03% Diol (8 tonnes).

2.5 Optics

SNO+ is an extremely sensitive probe of light created in its target volume. Relating these detected photons to the energy, position, time and type of particles producing them requires an understanding of the production, propagation and absorption of optical photons at the MeV scale. The next section examines each of these processes in turn, starting with the production of Cherenkov[2] and scintillation light.

2.5.1 Cherenkov Radiation

A charged particle travelling through a dielectric medium will emit Cherenkov radiation if its speed exceeds the local phase velocity of light. Photons are emitted from a shock in the polarisation field of the medium.

As a charged particle traverses the medium it polarises it. This is true at any speed, but the result is drastically different depending on whether the particle is faster or slower than the local light speed. The left hand diagram in Fig. 2.6 shows the case where the speed of the particle is slow compared with the speed of light. Here the polarisation field has time to compensate for the particle motion and remains

---

[2]the description of Cherenkov radiation roughly follows the one given in [17].

(a)   **(b)**

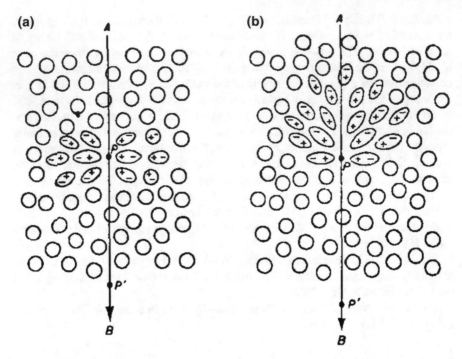

Fig. 2.6 Polarisation of a dielectric by a passing charged particle. Left: sub-luminal speed, right: super-luminal speed [17]

symmetric around the particle centre. Viewed at a distance, there is no overall dipole associated with the particle.[3] On the other hand, if the particle moves faster than the phase velocity of light (Fig. 2.6, right), the field does not have time to fully compensate for the motion of the particle and so, viewed instantaneously, there is an overall dipole along the direction of motion in the region around it. As the particle passes through the medium, sections of it gradually polarise and de-polarise by this mechanism and radiate as an oscillating dipole would, with intensity $\propto \sin^2 \theta$ w.r.t the dipole axis.

Adding up the wavelet contribution of each of the regions according to Huygens principle shows that light is emitted in a forward cone around the direction of particle motion, known as a Cherenkov cone, shown in Fig. 2.7.

The opening angle the cone, θ_C, for wavelength λ is fixed by the particle velocity v and the refractive index of the medium n

$$\cos \theta_C(\lambda) = \frac{1}{n(\lambda)} \frac{c}{v} \tag{2.1}$$

---

[3]Molecular sizes are 2–3 orders of magnitude larger than the wavelength of optical light and so dipoles of varying orientation average to zero.

Fig. 2.7 Huygens' principle
for Cherenkov wavelets [17]

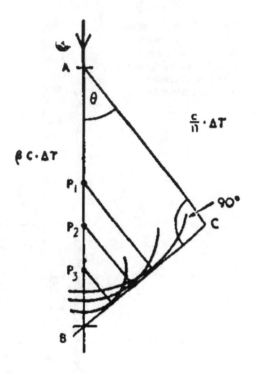

provided $v > c \cdot n(\lambda)$. As $v \to c$, θ_C converges on $\arccos(\frac{1}{n})$, which is $44°$ for optical
photons in LAB.

Emission begins immediately and continues until the particle velocity drops below
c/n. For 2.5 MeV electrons in LAB, the light is emitted within 60 ps (Fig. 2.8). This
is drastically less than the FWHM of the PMT TTS, so the emission is effectively
instantaneous.

The Frank–Tamm formula gives the number of emitted photons per unit track
length, N_γ, per unit energy loss energy E, produced by a particle with $\beta = v/c$
travelling though a material of refractive index n and permeability μ:

$$\frac{d^2 N_\gamma}{dx\, dE} = \frac{\alpha^2 Z^2}{\hbar c} \mu(E) \left(1 - \frac{1}{\beta^2 n(E)^2} \right) \tag{2.2}$$

This can be integrated over the length of the particle track with Monte Carlo.
Figure 2.8 shows just that for 2.5 MeV electrons in LAB; above 0.5 MeV, around
610 Cherenkov photons are produced per MeV of kinetic energy; below 1 MeV there
is significant non-linearity.

For both water and liquid scintillators $\mu(E)$, and $n(E)$ are approximately constant
in the optical range spectrum so the number of photons falls towards the red end of
the spectrum:

$$N(E) \sim \text{constant} \implies N(\lambda) \sim \frac{1}{\lambda^2} \tag{2.3}$$

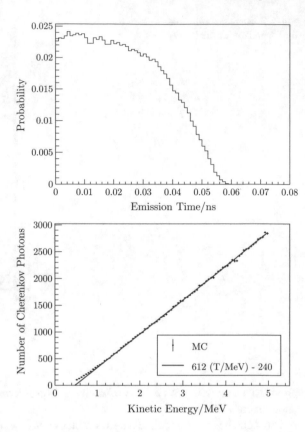

Fig. 2.8 Cherenkov photons in LAB. Top: emission times for 2.5 MeV electrons in LAB. Bottom: number of Cherenkov photons as a function of electron energy

2.5.2   Scintillation

Scintillation is an example of fluorescence: light emitted after a non-thermal excitation produced by ionizing radiation. The short (ns to μs) decay times of fluorescence distinguish it from, slower, phosphorescence (ms to s).

Organic liquids containing aromatic rings scintillate via the excitation and relaxation of highly de-localised electrons within those rings. The planar geometry of the ring leads to sp^2 hybridisation in the constituent carbons, whereby the states that overlap to form the primary σ-bonds between carbons are superpositions of the s orbital ($L = 0$) and two of the three p orbitals p_x, p_y ($L = 1$). This leaves the p_z orbitals, pointing perpendicular to the plane of the ring, to overlap and form secondary π-bonds. These π bonds are highly de-localised, so their excited states are typically separated by a few eV. Figure 2.9 shows an energy diagram of these π states.

Passing particles can excite electrons in these π orbitals either directly, via elastic scattering, or indirectly, via ionisation and recombination into an excited state. Once in one of these excited states, electrons radiatively de-excite and emit optical photons. The lifetime of this excited state and the wavelength of light emitted depends on the excitation's path through the energy levels in Fig. 2.9.

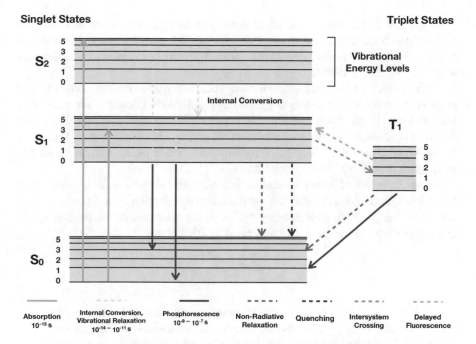

Fig. 2.9 Jablonski diagram of molecular π orbitals in a scintillating organic liquid

In the case of elastic scattering, electrons are almost always promoted to $S = 0$ singlet states, because transitions to $S = 1$ triplet states are forbidden by spin selection rules. Regardless of which singlet state the electron is excited to, it de-excites to the lowest excited state, S_{10}, within ps or less, by vibrational relaxation and internal conversion. From there, the most likely outcome is that the molecule decays to the ground state, or one of its vibrational excitations, with ns characteristic time, emitting primary scintillation photons (corresponding to $S_{10} \to S_{0X}$ in Fig. 2.9).

In order to be useful, a scintillator must have a Stokes' shift. This is the requirement that photons emitted by the scintillator are too low in energy to re-excite it, limiting self-absorption of the scintillation light. In organics the shift is provided by vibrational excitations: in Fig. 2.9, absorbed photons have energies of *at least* $S_{10}-S_{00}$, whereas emitted photons have energies of *at most* $S_{10}-S_{00}$.

Instead of scintillating, the excitation can return to the ground state without emission of a photon. This 'non-radiative' relaxation occurs directly, via overlap between S_{10} and vibrational excitations of the ground state $S_{01,2,3...}$. The ratio of the radiative and non-radiative outcomes determines the scintillator's fluorescence quantum yield: the fraction of excitations that lead to the emission of a photon.

Triplet excited states can be created indirectly from singlet states via inter-system crossing, but the process is relatively rare. More commonly, they are produced by ionisation-recombination, which produces triplets in 75% of cases. After vibrational de-excitation to T_0, the triplet state is very stable (again due to spin/parity selection). The radiative decay of these states over ms to tens of seconds is phosphorescence.

More relevant for scintillation detectors is delayed fluorescence, where the excitation is converted back into a singlet state, and the singlet state decays. If the decay is radiative, the result is emission that is identical to primary scintillation, but apparently with a longer time constant.

Light emitted on the time scale of ns to 100 ns is therefore of two components: primary fluorescence created by singlet decays and delayed fluorescence created by indirect triplet decay. The proportion of the latter component increases strongly with the ratio of ionisation-recombination to excitation by elastic scattering. This gives heavy ionising particles (e.g. α) a 'long tail' scintillation pulse that can be used to distinguish them from lighter particles (e.g. e^-) [13].

A second effect of heavy ionisation is ionization quenching. This is an effect whereby regions of dense ionization produce less scintillation light then those without ionization. A comprehensive description of the microscopic mechanism is still lacking, but the light yield is well described by Birk's law:

$$\frac{dL}{dx} = S \frac{\frac{dE}{dx}}{1 + k_B \frac{dE}{dx}} \tag{2.4}$$

here L is the light yield, x parametrises the particle's path, dE/dx is the particle's energy loss and k_B is Birk's constant, which is particle and scintillator dependent.

The performance of liquid scintillator detectors can often be improved by the inclusion of one or more dopants called fluors. Primary fluors are typically doped in concentrations of a few g/L. At these concentrations, optical photon absorption and molecular excitation are dominated by the scintillator itself, but excited scintillator molecules typically encounter a dopant molecule *before* they fluoresce, transferring the excitation to the dopant. Radiative transfer also occurs, but it is sub-dominant at these concentrations.

There are two primary advantages to transferring excitations to a dopant. First, the dopant may have a better quantum yield than the primary scintillator, even once the transfer efficiency between the two is taken into account. Second, the dopant will have a different emission profile to the primary scintillator, which can be chosen to enhance the Stokes' shift of the overall scintillator cocktail: if the primary fluor emits below the threshold energy for absorption by the primary scintillator, the scintillation light will not be re-absorbed.

In addition, one or more secondary fluors may be added at the mg/L level to radiatively shift the wavelength spectrum by absorption re-emission. Typically this is done to limit self-absorption by the primary fluor and to match the scintillator's emission spectrum to the optical response of the detector.

2.5.3 SNO+ Cocktail

The SNO+ LAB will be doped with 2g/L of the primary fluor 2,5-Diphenyloxazole (PPO). The quantum yield of PPO is 80% and non-radiative transfer from LAB to PPO is 75% efficient at this concentration.

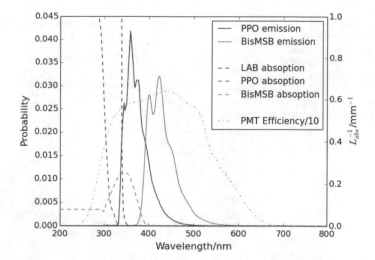

Fig. 2.10 Stokes' shift and optical matching for the SNO+ scintillator cocktail. The solid lines show emission spectra, the dashed lines show inverse absorption lengths and the dotted line shows the PMT combined efficiency curve [18–21]

During the tellurium phase, a second fluor, 1,4-Bis(2-methylstyryl)benzene (bisMSB), will be included at a concentration of 15 mg/L. Figure 2.10 shows how the two dopants enhance the scintillator's Stokes' shift: PPO emits light with wavelengths too long to be absorbed by LAB, and bisMSB emits at higher wavelengths than PPO is able to absorb. bisMSB also improves spectral overlap between the emitted light and the efficiency of the PMTs.

2.5.4 Attenuation and Re-emission

The absorption lengths of the SNO+ optical components are given in Fig. 2.11. For the wavelengths accepted by the PMTs, the scintillator attenuation lengths are long compared with the detector size.

Light absorbed by LAB has a 60% (= 80% × 75%) chance of being transferred to PPO and re-emitted. Light absorbed by PPO directly has an 80% chance of re-emission and light absorbed by bisMSB is re-emitted with 96% efficiency. The scattering length of LAB is over 10 m in the optical range [19, 22].

The scattering and absorption lengths of water are both of order 100 m in the optical range so optical attenuation in the external water is negligible. However, the mismatch between the refractive indices of the three components is significant: it causes reflections/refractions at the boundaries, particularly for events occurring close to the AV that tend to produce light that meets it at smaller angles.

Fig. 2.11 Absorption
lengths of SNO+ optical
components. The assumed
fluor concentrations are PPO
2g/L, bisMSB 15 mg/L
[20–22]. The horizontal lines
are conservative assumptions
where measurements are not
available

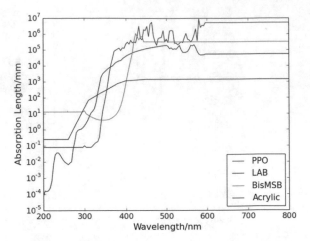

The AV thickness is only 55 mm so absorption by acrylic is only significant for
Cherenkov photons in the UV range.

2.5.5 Effect of Te Loading

TeDiol produces *chemical* quenching in the LABPPO mixture it is loaded into: at
0.5% loading, the light yield is reduced by 44%, to 6650 γ/MeV [18]. At this light
yield, the detector will collect 408 p.e./MeV, or 1020 p.e. at the $0\nu\beta\beta$ end point.

The quenching increases steeply with increasing concentration. The light level
determines the detector's energy resolution and, therefore, the rate of $2\nu\beta\beta$ mis-
reconstructions into the signal region of interest. The quenching thus limits the
amount of Te that can be usefully loaded into the scintillator. It has been explicitly demonstrated that the Diol has no effect on the emission *spectrum* of LABPPO
or LABPPO + bisMSB [18].

Figure 2.12 shows the effect of the TeDiol loading on LAB-PPO timing. In the
region dominated by the fastest decay time, the two curves are very similar. At later
times however, the longest time constant is preferentially quenched in the loaded
mixture; this will have a detrimental effect on α/β discrimination, which relies on
the prominent late tail of α scintillation pulses.

The absorption of the Diol itself has been measured explicitly across the relevant
wavelength range. It absorbs strongly in the UV range, but this light is otherwise
absorbed by LAB and is almost undetectable by the PMTs. More importantly, there
is close to zero absorption in the wavelength range emitted by bisMSB and PPO.

More recently, the possibility of loading dimethyldodecylamine (DDA) has
emerged. All evidence points towards DDA conferring stability to the TeDiol +
LABPPO cocktail and raising its light yield by as much as 10–15%. Although it is
likely that this will be deployed in SNO+, the work that follows uses simulations that
assume no DDA.

Fig. 2.12 Emission time distributions for LABPPO with and without 0.5% TeDiol loading [18]

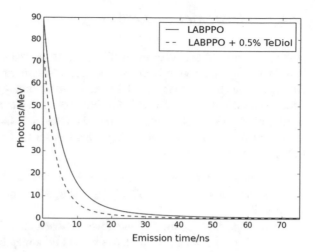

2.6 Detector Calibration

The account in this chapter so far gives a model for the processes that lead from a $0\nu\beta\beta$ event in the target volume to PMT hits built into analysable detector events. However, there are many free parameters in this model that are a priori unknown. Constraining these parameters is the task of calibrations.

Understanding the SNO+ scintillator requires knowledge of its light yield, the timing and wavelength distributions of the light it emits, the attenuation and scattering coefficients that govern light propagation through it, and the optical matching where light leaves it.

Equally important is understanding the PMTs, their concentrators and their read-out channels. The detector collects events comprised of digitized times and charges, read out from the PMTs. These readout times must be related to the arrival times of the incident photons. This is difficult because each PMT produces a differently shaped charge pulse, each TAC behaves differently and issued GTs arrive at different times for each channel. In addition, the the digitized charges must be related to the charge deposited at the anode of each PMT. Each channel has a different dark noise rate, a different threshold, and each integrator has a different charge pedestal. Finally, the reflectivity of the concentrators and its variation with incidence angle need to be measured explicitly.

Individual PMTs and small scintillator samples can be measured on the bench-top but optical properties have been demonstrated to vary with volume scale [23] and the PMT on the bench-top sits in a different environment to those in the detector. Hence the need to measure these properties directly in the detector with in-situ calibration.

2.6.1 Electronics Calibration

Front end calibrations focus on the PMTs and their readout channels. For this, there are two calibration routines, ECAL and ECA, that are run directly on the electronics using features built into the DAQ. ECAL performs many on-board diagnostic procedures, and sets the discriminator thresholds of each channel to a level that results in dark hits at 500 hz. ECA runs calibrate the time and charge response of each channel using forced discriminator firing (PED) and forced global triggering (PULGT). There are two types of runs:

- **Charge pedestals** run PED on a channel, followed by a PULGT 150 ns later. These ECA runs measure the DC offset in the charge integrators channel by channel, cell by cell.
- **TAC Slope** By varying the time offset between the PED and the PULGT in 10 ns steps (0, 10, 20, 30…500 ns), these runs build a map between TAC values and hit time before trigger. This is performed channel by channel, and for both TACs on each channel.

During data processing, ECA calibrations are applied to every hit: pedestals are subtracted for each integrated charge, and the TAC is converted into a time relative to the GT time using the measured slopes.

2.6.2 Optical Calibrations

Optical calibrations use two well understood light sources to calibrate the PMT response and optical properties of the detector components. The laserball is deployed directly into the scintillator, while the ELLIE system injects light into the detector via optical fibres mounted on the PSUP.

The laserball [24] is a quartz sphere, 10.9 cm in diameter, filled with 50 μm diameter air filled glass beads suspended in soft silicone gel. A nitrogen laser on the deck above the detector is passed through one of several dye resonators to produce light at several optical wavelengths. Fibre optic cables feed this laser light into the laserball to produce near isotropic light. The laser intensity is tuned to produce only single p.e. hits on any PMT and, by forcing the detector to trigger asynchronously, the laserball produces light with a fixed, known relationship to the GT. The laserball is deployed into the AV and moved through the target volume using a system of ropes controlled from the deck. The laserball data are used to fit an optical model, calibrating the PMTs themselves and extracting optical parameters [24]; its major elements are:

- **Cable delay** Comparing the global offsets between channels measures the GT delay per channel.
- **Time walk** The discriminator crossing time of the channels is charge dependent, mapping out the charge vs. hit time calibrates this effect.
- **PMT efficiency** Comparing PMT hit rates allows for relative channel by channel efficiency measurements.

- **Attenuation lengths/angular response** Correlating across different laserball positions allows the attenuation lengths of the target, AV and external water to be extracted. The angular response of the concentrators is extracted from the angular variation of the PMT hit probability.

During data processing, a correction is made to each PMT hit time using its measured cable delay and time walk effect.

The ELLIE (Embedded LED/Laser Light Injection Entity) is designed to allow in-situ optical calibrations without the dead-time and radio-contamination risk of deploying the laserball. Light is injected into the detector from inward pointing obtical fibres, mounted onto the PSUP.

Its timing system, TELLIE [25], consists of 92 fibres and LED drivers. The LEDs produce light pulses of 10^{3-5} photons in under 5 ns, and the fibres distribute the photons into a wide cones of $20°$ half angle, allowing redundant cover of all inward facing PMTs. Like the laserball, it can asynchronously trigger the detector and the intensity is chosen to produce single p.e. hits. From these runs, the cable delays and time walk of the PMTs can be measured [25].

The scattering module, SMELLIE [26], measures the Rayleigh scattering length and scattering angle of the cocktail, using laser light injected into one of 15 fibres across the PSUP, directed across the detector at different angles, and collimated into narrow $3°$ beams. The light is produced by one of 4 fixed wavelength lasers or a super continuum laser that produces light in the range $\lambda = 400-700$ nm. The scattering angle and length of the scintillator cocktail can be measured using analysis of the hit probability and charge distributions as a function of angle w.r.t fibre direction [27].

2.6.3 Radioactive Sources

Once light propagation and detection is understood, the final piece of the puzzle is the behaviour of primary particles. For this purpose, several known radioactive sources have been prepared for deployment into the detector; a list of sources under development for the scintillator phases is given in Table 2.1. Key measurements include the energy scale of the detector (the overall number of hits per MeV of deposited energy) and the performance of position/energy fitters.

Table 2.1 Radioactive sources for SNO+ scintillator phases [28]. The γ response will differ from the electron response. The difference between them can be estimated with Monte Carlo and calibrated using in-situ sources

Isotope	Particles	Dominant mode visible energy/MeV	Calibrates
AmBe	n, γ	(2.2 + 4.4)	Neutron capture response
^{46}Sc	γ	1.1 + 1.3	Energy scale & reconstruction
^{57}Co	γ	0.1	Energy scale & reconstruction
^{48}Sc	γ	1.0 + 1.1 + 1.3	Energy scale & reconstruction
^{8}Li	Optical photons	–	Cherenkov light

2.7   RAT Simulation

The RAT (Reactor Analysis Tools) software package was originally written by Stan Siebert for the Braidwood collaboration. The SNO+ version of RAT serves as the event generator, detector simulation and on-line analysis package for the experiment.

It contains a Geant4 based simulation of the full detector geometry, including all standard electro-magnetic and hadronic physics relevant at the MeV scale. In addition, the GLG4sim package, originally produced for the KamLAND experiment, is used for primary event generation and the simulation of scintillation photons. Every optical photon is tracked through the detector, accounting for reflections, refraction, scattering and absorption. Events from radioactivity (including $0\nu\beta\beta$) are simulated using a C++ port of the Decay0 generator written by V. Tretyak [29].

The PMTs and concentrators are modelled as full 3D objects. The front end and trigger system are simulated in full, down to each discriminator and each trigger signal. Artificial electronics noise is added to each PMT hit during 'uncalibration' and corrected for using calibration constants measured on the detector.

After running event by event analysis on the produced Monte Carlo/data, the results are written to disk as ROOT files for analysis.

References

1. Boger J et al (2000) The Sudbury neutrino observatory 449:172–207
2. Beltran B et al (2013) Measurement of the cosmic ray and neutrino-induced muon flux at the sudbury neutrino observatory 01:1–17. arXiv:0902.2776v1
3. Eguchi K et al (2003) First results from KamLAND : evidence for reactor antineutrino disappearance (January):1–6. https://doi.org/10.1103/PhysRevLett.90.021802
4. Fukuda S et al (2003) The Super-Kamiokande detector. Nucl Instrum Methods Phys Res Sect A Accel Spectrometers Detect Assoc Equip 501(2):418–462. https://doi.org/10.1016/S0168-9002(03)00425-X, http://www.sciencedirect.com/science/article/pii/S016890020300425X
5. Alimonti G et al (2008) The Borexino detector at the Laboratori Nazionali del Gran Sasso (June). arXiv:0806.2400v1
6. Alimonti G et al (2009) Nuclear instruments and methods in physics research a the liquid handling systems for the Borexino solar neutrino detector 609:58–78. https://doi.org/10.1016/j.nima.2009.07.028
7. Andringa S et al (2016) Current status and future prospects of the SNO + Experiment. arXiv:1508.05759v3 [physics.ins-det]. Accessed 28 Jan 2016
8. Caden E (2017) Private communication
9. Lay MD, Lyon MJ (1996) An experimental and Monte Carlo investigation of the R1408 Hamamatsu 8-inch photomultiplier tube and associated concentrator to be used in the Sudbury Neutrino Observatory. Nucl Instrum Methods Phys Res Sect A Accel Spectrometers Detect Assoc Equip 383(23):495–505. https://doi.org/10.1016/S0168-9002(96)00861-3, http://www.sciencedirect.com/science/article/pii/S0168900296008613
10. Biller SD et al (1999) Measurements of photomultiplier single photon counting efficiency for the Sudbury Neutrino Observatory. Nucl Instrum Methods Phys Res Sect A Accel Spectrometers Detect Assoc Equip 432(2):364–373. https://doi.org/10.1016/S0168-9002(99)00500-8, http://www.sciencedirect.com/science/article/pii/S0168900299005008
11. Heintzelman W, Private communication, Private communication

12. Wan Chan Tseung Ã H, Kaspar J, Tolich N (2011) Nuclear instruments and methods in physics research a measurement of the dependence of the light yields of linear alkylbenzene-based and EJ-301 scintillators on electron energy. Nucl Inst Methods Phys Res A 654(1):318–323. https://doi.org/10.1016/j.nima.2011.06.095, http://dx.doi.org/10.1016/j.nima.2011.06.095

13. Chen MC (2011) Scintillation decay time and pulse shape discrimination in oxygenated and deoxygenated solutions of linear alkylbenzene for the SNO + experiment 00:1–5. arXiv:1102.0797v1

14. Ford RJ (2015) A scintillator purification plant and fluid handling system for SNO+. In: American institute of physics conference series. American Institute of Physics Conference Series, vol 1672, p 080003. https://doi.org/10.1063/1.4927998, arXiv:1506.08746

15. Caleb Miller (2016) Analysis of cosmogenic impurities in tellurium and development of tellurium loading for the SNO+ experiment

16. Wright A (2017) Te Dev R&D Pre-summary, SNO+-docDB 4545-v1

17. Jelley JV (1961) Cerenkov radiation: its origin, properties and applications. Contemp Phys 3(1). https://doi.org/10.1080/00107516108204445

18. Mastbaum A, Barros N, Coulter I, Kaptanoglu T, Segui L. Optics overview and proposed changes to RAT, SNO+-docDB 3461

19. Segui L. Scintillator model: comparison between new data and old model, SNO+-docDB 2774

20. Dai X. Te-diol tests at Queen's, SNO+-docDB 3315

21. Segui L. Te-diol studies: stability and optics, SNO+-docDB 3880

22. Liu Y. Attenuation and scattering of TeBD & the cocktail, SNO+-docDB 3880

23. Suekane F et al (2004) An overview of the KamLAND 1-kiloton liquid scintillator. ArXiv Physics e-prints, arXiv:physics/0404071

24. Stainforth R. Characterising the optical response of the SNO+ detector. PhD thesis

25. Alves R et al (2015) (SNO+), The calibration system for the photomultiplier array of the SNO+ experiment. JINST 10(03):P03002. https://doi.org/10.1088/1748-0221/10/03/P03002, arXiv:1411.4830

26. Gagnon N, Jones C, Lidgard J, Majumdar K, Reichold A, Segui L, Clark K, Coulter I. The SMELLIE hardware manual, SNO+-docDB 3511-v2

27. Majumdar K. On the measurement of optical scattering and studies of background rejection in the SNO+ detector. PhD thesis

28. Peeters S. Source and interface list, SNO+-docDB 1308-v9

29. Ponkratenko OA, Tretyak VI, Zdesenko YuG (2000) The Event generator DECAY4 for simulation of double beta processes and decay of radioactive nuclei. Phys Atom Nucl 63:1282–1287. https://doi.org/10.1134/1.855784, [Yad. Fiz. 63,1355(2000)], arXiv:nucl-ex/0104018

Chapter 3
Reconstruction

In order to turn the raw hit and trigger data read out in SNO+ detector events into a $0\nu\beta\beta$ analysis, one must first determine what events produced them. Reconstruction is the process of inferring high level particle information from the data read out in each event.

ScintFitter is the SNO+ reconstruction algorithm for the scintillator and tellurium phases, implemented in RAT. It is important to describe here because it is used to produce the vertex positions used to classify events in Chaps. 5 and 6, the signal extraction PDFs used in Chap. 8, and the seed for the reconstruction algorithms developed in Chap. 9. The first half of this chapter describes ScintFitter itself; this is mostly a necessary description of other's work but it pays particular attention to a multi-hit correction method developed by the author. The second half of the chapter introduces the ideas that will be applied in Chaps. 5 and 6; it shows that each event's topology is encoded in the PMT hit times it produces and sets out some general principles for event classification using those times.

3.1 ScintFitter

SNO+ reconstruction relies on the PMT hit times to determine event time and position. The true event time determines a global offset on the hit times, whereas the event position determines the shape of the hit time distribution. For example, events at the detector centre produce photons that all arrive at approximately the same time, but events near the AV create photons with a range of arrival times: the PMTs closest to the event are hit first and those on the far side of the detector hit last.

The event energy is determined from the number of PMT hits. Higher energy particles deposit more energy in the scintillator, produce more scintillation photons and trigger more PMT hits.

© Springer Nature Switzerland AG 2019
J. Dunger, *Event Classification in Liquid Scintillator Using PMT Hit Patterns*,
Springer Theses, https://doi.org/10.1007/978-3-030-31616-7_3

`ScintFitter` reconstructs the time, position and energy of events under the assumption that they are 3 MeV electrons. The fit proceeds in three stages: first, a simple seed time and position is estimated using `QuadFitter`; second, a more accurate time and position is estimated using `PositionTimeLikelihood`; and, finally, the energy is estimated using the number of hits in the event (N_{hit}). The next sections describe each of these methods in turn, summarising their overall performance.

3.1.1   Time and Position Fit

`ScintFitter` estimates the event time relative to trigger and the position of the event vertex in detector coordinates. These detector coordinates are Cartesian coordinates with respect to an origin at the centre of the AV. Alongside these, it is useful to define the event radius, $r = \sqrt{x^2 + y^2 + z^2}$, and the polar angle, θ, defined by the smallest angle a vector makes with the z-axis, which points towards the neck.

The position and time fit proceeds in two stages. `QuadFitter` provides a rough estimate of the vertex using an analytical calculation, then `PositionLimeLike lihood` improves this estimate with a likelihood fit, using the `QuadFitter` result as a seed.

For a uniform detector with perfect timing resolution, the event time and position of a instantaneous, point-like event is exactly calculable using four PMT hits and the time of flight equation:

$$|\vec{x}_h - \vec{x}_v| = c(t_h - t_v) \tag{3.1}$$

here $\vec{x}_h$, t_h are the position and time of a PMT hit, c is the photon speed in the detector material and $\vec{x}_v$, t_v are the event time and position. Four PMT hits produce four equations, which can be used to determine the four unknown parameters, x_v, y_v, z_v, t, without uncertainty.

`QuadFitter`, developed by Coulter and others [1], adapts this principle for the imperfect SNO+ detector. Noise hits, PMT jitter and the broad scintillation timing profile mean that the solution to Eq. 3.1 will depend on the 4 PMT hits chosen. `QuadFitter` takes a large number of sets of 4 PMTs, calculates x_v, y_v, z_v, t_v for each set and then selects the median of each parameter as the best fit. The effective speed in Eq. 3.1 is chosen to minimise the radial bias of events in the central 5 m.

`PositionTimeLikelihood`, written by Jones and Coulter [1, 2], takes the `QuadFitter` result as a seed and improves on its estimate in two ways. First, the single `QuadFitter` light speed is adapted to take into account the different light speeds in acrylic, water and scintillator. Second, the distributed emission of the scintillator is modelled using an emission time PDF and the method of maximum likelihood.

`PositionTimeLikelihood` extends Eq. 3.1 by calculating the path lengths in water, scintillator and acrylic for each hit. It assumes straight line paths and converts to time of flight using effective velocities for each material:

Fig. 3.1 Time residual PMT hit PDF used in the ScintFitter position fit

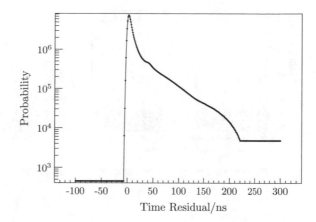

$$t_{t.o.f} = \frac{d_w}{c_w^{eff}} + \frac{d_a}{c_a^{eff}} + \frac{d_s}{c_s^{eff}} \tag{3.2}$$

where $d_{s,a,w}$ and $c_{s,a,w}^{eff}$ are the path lengths and effective speeds in scintillator, acrylic and water, respectively.

To account for distributed emission, the apparent emission times after the event start are calculated using 'time residuals':

$$t_{res}(\vec{x}_v, t_v) = t_h - t_{t.o.f} - t_v \tag{3.3}$$

The time residuals for a hypothesised vertex can be compared with the distribution that would be observed using the true vertex, $P(t_{res})$. The best fit vertex time and position are chosen to maximise agreement with this distribution, using the method of maximum likelihood and Powell optimisation [3]. Fits that do not converge are flagged as invalid. The log-likelihood is:

$$\log \mathcal{L}(\vec{x}_{ev}, t_{ev}|h^i) = \sum_{i=0}^{N_{hits}} \log P(t_{res}^i(\vec{x}_{ev}, t_{ev})) \tag{3.4}$$

Figure 3.1 shows the PDF used in PositionTimeLikelihood; its shape is dominated by the scintillator emission profile (Fig. 2.12), but it differs in the presence of noise hits, smearing from the PMT TTS, and late hits from scattered, re-emitted and reflected light.

Figure 3.2 summaries the ScintFitter position fitter performance on $0\nu\beta\beta$ events occurring in the central 3.5 m of the detector. d is the projection of the vector that points from the truth vertex onto the true electron direction. $\|\vec{x}_{fit} - \vec{x}_{truth}\|$ shows the absolute distance between the true and reconstructed position. The x, y, z, t, r plots show the difference between the reconstructed coordinates and their truth equivalents. The x, y, z coordinates are estimated with a Gaussian resolution of 5.7 cm

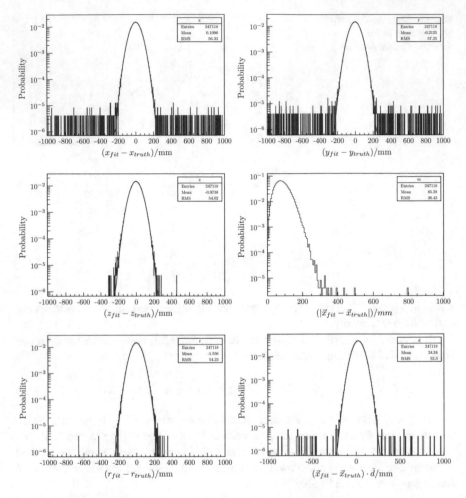

Fig. 3.2 Fit performance for $0\nu\beta\beta$ events $r/\mathrm{m} < 3.5$ for valid fits with the ITR cut described in Sect. 3.2.1. x, y, z are Cartesian positions co-ordinates w.r.t the detector centre. $\hat{d}$ is the true direction of the first electron (arbitrarily chosen), $r = \sqrt{x^2 + y^2 + z^2}$

and negligible bias but there is a 2.5 cm bias along the direction of motion of the electron. Overall, the true vertex and the reconstructed vertex are separated by 8.5 cm on average.

3.1.2 Energy Fit

At MeV energies, the number of scintillation photons produced in an event is proportional to the (quenched) energy deposited in the scintillator, E_{dep}^Q. The *expected* number of p.e. collected across the detector, $\overline{N}_{p.e.}$, will also scale proportionally:

$$\overline{N}_{p.e.} \propto E_{dep}^{Q} \qquad (3.5)$$

but, in any given event, there are Poisson fluctuations in the number of p.e. actually collected, $N_{p.e.}$:

$$N_{p.e.} \sim \mathrm{Poi}(\overline{N}_{p.e.}) \qquad (3.6)$$

For large $\overline{N}_{p.e.}$, this is approximately a Gaussian with width $\sigma_{\mathrm{Poi}} = \sqrt{N_{p.e.}}$. σ_{Poi} places a fundamental limit on the resolution of *any* energy reconstruction algorithm: even if the constant of proportionality in Eq. 3.5 is known exactly, the resolution is limited by the statistical width of the $N_{p.e.}$ distribution. This optimal resolution is referred to as the Poisson limit.

The SNO+ detector will collect 1020 p.e.[1] at $Q_{\beta\beta}$, so the fractional energy resolution there is limited to $1/\sqrt{1020} = 3.1\%$, or 78 keV. Even deviations of a few % from this limit can damage $0\nu\beta\beta$ sensitivity, because the number of background $2\nu\beta\beta$ counts in the signal region is very non-linear in energy resolution, $n_{2\nu} \propto \sigma_E^{5.5}$ (1.7.5).

This simple picture is made more complicated by the fact that SNO+ measures N_{hit}, not $N_{p.e.}$, and because the constant of proportionality in Eq. 3.5 depends on vertex position. ScintFitter accounts for the first effect using a multi-hit correction and the second using a functional form that relates the number of p.e. generated by events at the detector centre to the number of p.e. generated at an arbitrary position. These two corrections are described in detail in the following sections.

Multi-hits

The first complication comes from p.e. lost to the pile up of multiple hits on a single PMT. In any physics event, a large number of photons are created, each with a very small probability of causing a hit in the ith PMT. The result is a Poisson distribution in the number of p.e. collected in i:

$$N_{p.e.}^{i} \sim \mathrm{Poi}(\mu^{i}) \qquad (3.7)$$

where μ^{i} is the expected number of p.e. in this PMT for this event. The probability of registering a hit is then approximately the probability of observing one or more photo-electrons[2]:

$$P_{hit} = P(N_{p.e.} \geq 1) = 1 - \exp(-\mu) \qquad (3.8)$$

---

[1]For events with $r_{fit} < 1$ m. The total number of generated p.e. is around 20% larger, but not all PMTs receiving a p.e. will cross threshold.

[2]The relation won't hold exactly because the p.e. are not completely independent: a PMT is more likely to register a hit if it has already recieved a p.e. that did not cause it to cross charge threshold.

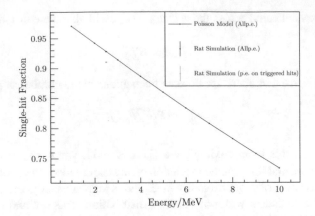

Fig. 3.3 The fraction of p.e. that hit a PMT with no other p.e., as a function of event energy in the SNO+ Te phase I, calculated using a simple Poisson model. The red line shows the fraction for $0\nu\beta\beta$ events at 2.5 MeV, $r < 1$ m, estimated using RAT6.1.6. The blue point shows the single hit fraction for all p.e. (i.e. including p.e. on PMTs that do not cross threshold), the green point shows the same fraction including only those p.e. that cross threshold, causing a hit. The model agrees perfectly when all p.e. are considered, but over-estimates the single p.e. fraction when only the hit PMTs are considered. This is because a multi-hit PMT is more likely to cross threshold than one with a single p.e. The calculation assumes there are 9023 on-line, inward PMTs, that the light is isotropic, and that there are 532.2 p.e./MeV on those 9023 PMTs

and the fraction of single p.e. hits is

$$F = \frac{P(N_{p.e} = 1)}{P(N_{p.e.} \geq 1)} = \frac{\mu \exp(-\mu)}{1 - \exp(-\mu)} \tag{3.9}$$

Figure 3.3 shows the single p.e. fraction as a function of event energy calculated using Eq. 3.9. At 2.5 MeV, around 7% of p.e. are lost to multi-hits. This difference is significant: it corresponds to a reduction in energy resolution of 3.5 and 29% more $2\nu\beta\beta$ background in the region of interest. At 10 MeV the fraction of p.e. lost is over 25%. The effect is more severe further away from the centre of the detector.

Multi-hit Correction

A simple solution would be to infer the number of p.e. from the total charge collected by each PMT, however the charge collected by the SNO+ PMTs is a very weak estimator of $N_{p.e.}$ at low hit multiplicity. The alternative, presented here, is to group the PMTs into ensembles that are approximately equally illuminated and then infer $N_{p.e.}$ from the statistical properties of those ensembles.

The crux of the problem is to estimate the number of p.e. in the ith PMT, $N_{p.e.}^i$, when the detector only records whether the PMT is hit or not, $h^i = (0 \text{ or } 1)$. On its own, Eq. 3.8 isn't very useful: observing a hit on PMT i tells you no more about P_{hit}

than a single head tells you the fairness of a coin. However, with N_{obs} observations of h^i, P^i_{hit} can be estimated from the number of hits on the PMT, N^i_{hit}:

$$\hat{P}^i = \frac{N^i_{hit}}{N_{obs}} \tag{3.10}$$

Physics events cannot be engineered to happen twice. But, in a single event, one can choose a set of PMTs with approximately the same illumination and therefore approximately the same $\mu^i = \mu$. Then, the expected number of hits in the ensemble is:

$$\overline{N}_{hit} = \sum_{i=0}^{N_{PMT}} (1 - \exp(-\mu^i)) = N_{pmt}(1 - \exp(-\mu)) \tag{3.11}$$

where N_{PMT} is the number of PMTs in the ensemble. Rearranging for μ gives:

$$\mu = -\log\left(1 - \frac{\overline{N}_{hit}}{N_{pmt}}\right) \tag{3.12}$$

Of course, in reality, no two PMTs share exactly the same μ because each PMT subtends a different solid angle with respect to the event vertex and PMT efficiencies vary etc. However, it should be possible to choose a set of PMTs with *approximately* the same μ:

$$\mu^i = \tilde{\mu} - \epsilon^i \tag{3.13}$$

for some central value $\tilde{\mu}$ for which $\epsilon^i << \tilde{\mu}$. In that case, the expected number of hits is given by:

$$\overline{N}_{hit} = \sum_{i=0}^{N_{PMT}} (1 - \exp(-\tilde{\mu} + \epsilon^i)) \tag{3.14}$$

$$= N_{PMT} - \exp(\tilde{\mu}) \sum_{i=0}^{N_{PMT}} \exp(\epsilon^i) \tag{3.15}$$

Equation 3.15 matches the expression for the ensemble with identical PMTs (Eq. 3.12), provided:

$$\frac{1}{N_{PMT}} \sum_{i=0}^{N_{PMT}} \exp(\epsilon^i) = 1 \tag{3.16}$$

If the variation between the PMTs is small, $\epsilon << 1$, then this condition becomes:

$$\sum_{0}^{N_{PMT}} \epsilon^i = 0 \tag{3.17}$$

if this condition is satisfied,[3] the representative $\tilde{\mu}$ is just the arithmetic mean of the μ of each PMT, $\tilde{\mu} = \bar{\mu}$, because:

$$\sum_{i=0}^{N_{hit}} \epsilon^i = \sum_{i=0}^{N_{hit}} \mu^i - \tilde{\mu} = \sum_{i=0}^{N_{hit}} \mu^i - \bar{\mu} = N\bar{\mu} - N\bar{\mu} = 0 \tag{3.18}$$

So, provided variations are small, the ensemble behaves although every PMT has $\mu^i = \bar{\mu}$ and the total number of p.e. in the ensemble is given by:

$$N_{p.e.}^{ens} = -N_{PMT} \log\left(1 - \frac{\overline{N}_{hit}}{N_{PMT}}\right) \tag{3.19}$$

The PMTs can be grouped into as many ensembles as required to have $\epsilon^i << \bar{\mu}$ hold everywhere. Then the total number of p.e. collected by the detector is well approximated by the sum over the ensembles:

$$N_{p.e.}^{detector} = -\sum_{ens} N_{PMT}^i \log\left(1 - \frac{\overline{N}_{hit}^i}{N_{PMT}^i}\right) \tag{3.20}$$

where N_{PMT}^i and $\overline{N}_{hit}^i$ are the number of PMTs, and the expected N_{hit} in ensemble i, respectively. This derived relationship can be used to produce an energy estimator, the multi-hit corrected hit count, N_{hit}^{corr}:

$$N_{hit}^{corr} = -\sum_{ens} N_{PMT}^i \log\left(1 - \frac{N_{hit}^i}{N_{PMT}^i}\right) \tag{3.21}$$

The critical difference between Eqs. 3.21 and 3.20 is that the expected hit count in each ensemble, $\overline{N}_{hit}$, has been replaced by the number actually observed, N_{hit}, introducing statistical variations. For low energy events, N_{hit}^{corr} is equivalent to the total N_{hit}:

$$N_{hit}^{corr} \rightarrow N_{hit} \text{ as } \frac{N_{hit}^i}{N_{PMT}^i} \rightarrow 0 \tag{3.22}$$

but at high energies, it corrects the N_{hit} upwards to account for multi-hits.

At very high hit probabilities, there is a significant pathology in N_{hit}^{corr}: if all of the PMTs in an ensemble are hit, the estimate for the hit probability is $\hat{P} = 1$, from which Eq. 3.21 implies an infinite light output $\hat{\mu} \rightarrow 1$. In practice, the PMT ensembles must be chosen to balance the uniform illumination condition with the variance of $\hat{P}$ and the risk of $\hat{P} = 1$ estimates.

---

[3]Note that the condition is absolute, rather than a comparison with $\tilde{\mu}$, this means that variation between PMT efficiencies sets an upper limit on μ for this technique. The method breaks down once $\mu > 1/\Delta E$, where ΔE is the maximum *fractional* difference between any two PMT efficiencies within the ensemble.

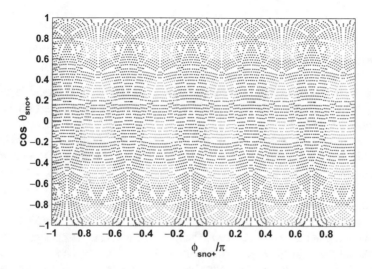

Fig. 3.4 PMTs divided into 100 equal solid angle segments about the detector centre

Fig. 3.5 N_{hit} and N_{hit}^{corr} as functions of energy for electrons at the centre of the SNO+ detector in the pure scintillator phase (the scintillator phase will have a higher light yield than the tellurium phase)

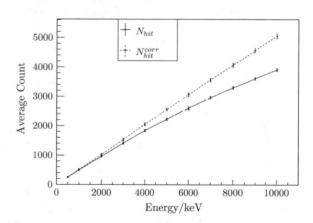

Segmentor

A new Segmentor class was added to RAT to divide the PMTs into these ensembles. The 4π solid angle, as viewed from any given point in the detector, is divided in to n_{div}^2 equal solid angle segments by dividing the regions $\cos\theta \in [-1, 1]$ and $\phi \in [-\pi, \pi]$ into n_{divs} intervals of equal length.

Figure 3.4 shows a flat map of the PMTs divided into 10×10 such equal solid angle segments about the detector centre. Figure 3.5 shows the comparative performance of N_{hits} and N_{hits}^{corr} calculated using these ensembles, over the range 0–10 MeV. N_{hit} begins to saturate above 1 MeV, whereas the corrected statistic remains linear well into the multi-hit regime. This linear behaviour leads to smaller errors propagated into the energy estimate from the observed count.

Asymmetry Corrections

To get close to the Poisson limit, reconstruction must also account for the dependence of $N_{p.e.}$ on event position. This is dominated by two asymmetries, one radial and one in the polar angle θ.

As the event vertex moves away from the detector centre, the number of hits tends to increase, because the increased solid angle and reduced attenuation of the photons travelling the short length across the detector more than compensates for those going the other way. Close to the AV surface itself however, very acute incident angles can produce total internal reflection that greatly reduces the number of hits.

The asymmetry in θ arises because the PMTs are not uniformly distributed in solid angle about the centre: there are fewer PMTs in the northern hemisphere because of the neck and the rope systems are not symmetric about the equator.

The `EnergyFunctional` method of `ScintFitter` fitter corrects for these effects using empirical functions which relate the expected N_{hits}^{corr} at arbitrary (r, θ) to events of the same energy at the detector centre $N_{hits}^{corr}(0, -)$:

$$\overline{N}_{hits}^{corr}(r, \theta) = \overline{N}_{hits}^{corr}(0, -)f(r|\theta) \tag{3.23}$$

$f(r|\theta)$ is continuous polynomial function of radius, piece-wise in θ, estimated using Monte Carlo.

`ScintFitter` fits the energy of a particle after first estimating its time and position. The N_{hits}^{corr} is first calculated according to Eq. 3.21 using 100 segments calculated around the detector centre. This is then related to an equivalent N_{hit}^{corr} deposit at the detector centre using Eq. 3.23 and compared with a look-up table that relates energy deposited to N_{hits}^{corr} observed at the centre. Figure 3.6 shows the performance of `ScintFitter` energy reconstruction on $0\nu\beta\beta$ events. The bias is statistically significant but negligible compared with the resolution. The energy resolution of $\sigma = 82$ keV is only 5% greater than the Poisson limit of 78 keV for 403 p.e./MeV.

3.2 Event Topology with Time Residuals

Once the event is reconstructed, it is useful to calculate the time residuals with respect to the fit vertex:

$$t_{res} = t_{hit} - t_{t.o.f} - t_{fit} \tag{3.24}$$

where $t_{t.o.f}$ is calculated from $\vec{x}_{fit}$ to $\vec{x}_{pmt}$ using Eq. 3.2.

These 'reconstructed time residuals' are an estimate of the photon emission times for a given event *assuming it's an electron*. Figure 3.7 shows these for 2.5 MeV

Fig. 3.6 `ScintFitter` energy reconstruction performance on $0\nu\beta\beta$ events $r < 3.5$ m. Valid fits, $r < 3500$ mm, with ITR cut

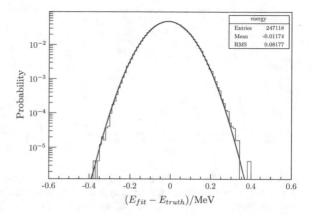

Fig. 3.7 Time residuals for 2.5 MeV electrons, generated in the AV using the reconstructed position

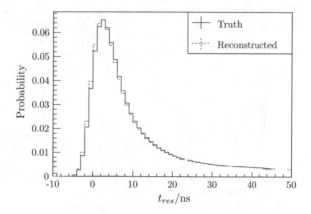

electron events generated uniformly in the AV, alongside the 'MC time residuals', calculated with respect to the true position and time. Comparing the curves reveals that finite vertex resolution leads to a slight broadening of the prompt peak rising edge.

The shape of the time residuals is dependent on the event energy. Figure 3.8 shows that the reconstructed time residuals for electrons, for several energy bins in the range 1–5 MeV. The distributions become more strongly peaked as the energy increases because more hits means smaller vertex resolution and because of the increasing number of multi-p.e. hits which arrive earlier on average.[4]

---

[4]The first of two photons will, on average, arrive before a single photon.

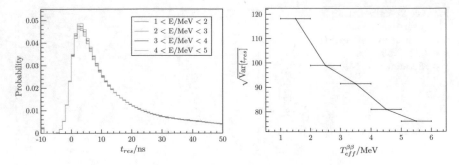

Fig. 3.8 The dependence of the electron time residual spectrum on reconstructed energy, $r_{fit} <$ 3.5 m. Left: the time residual spectra. Right: the standard deviation of the distributions. The events were generated uniformly in the AV, and uniformly in $1\,\text{MeV} < E < 10\,\text{MeV}$

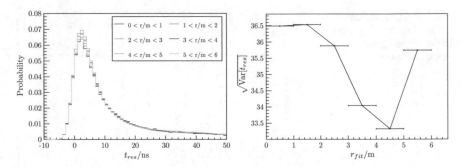

Fig. 3.9 The dependence of the electron time residual spectrum on reconstructed radius, for 2.5 MeV electrons. Left: the time residual spectra. Right: the standard deviation of the distributions. The events were generated uniformly in the AV

Reconstructed time residuals also run with event radius. Figure 3.9 shows that the spectra become more strongly peaked as the event radius increases, until the near AV region ($r > 5\,\text{m}$), where they become broader again. This must be a physical effect rather than a reconstruction effect, because the same trend is shown in the MC residuals (not shown here). Moving from the detector centre to the edge of the near AV region, absorption preferentially selects short paths which are more likely to be straight and, therefore, more likely to be properly accounted for by the straight line $t_{t.o.f}$ calculation. However, in the near AV region itself, total internal reflection effects lead to very poor time of flight corrections, and the trend is reversed.

There are two reasons an event's time residuals might not look like those in Fig. 3.7: either the vertex is reconstructed badly, or the event was not an electron. The first effect can be used to reject bad fits (ITR classifier) the second may be used to classify event types.

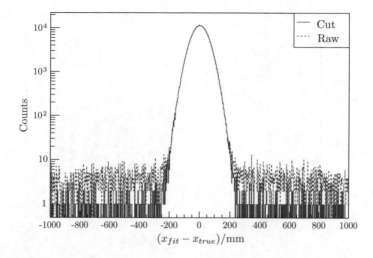

Fig. 3.10 Valid position fits for $0\nu\beta\beta$ events before and after the cut $0.33 < \text{ITR} < 0.45$, $r < 3500\,\text{mm}$

3.2.1 The ITR Classifier

The In Time Ratio classifier (ITR onwards) uses the reconstructed time residuals as a figure of merit for the position fit. To test consistency with Fig. 3.7, the fraction of hits with $-2.5 < t_{res}/\text{ns} < 5$ is compared with a reference window of $0.33 < \text{ITR} < 0.45$. Events with ITR values outside of this window are rejected as bad fits. Figure 3.10 shows the effect on the x position fits for $0\nu\beta\beta$ events which reconstruct inside the central 3.5 m, the non-Gaussian tails of the distribution are reduced by roughly a factor of two.

3.2.2 Timing Signatures for PID

Complex event topology is the second cause of time residual distortion. The time residuals considered so far are for single vertex electron events. If an event deposits energy in several vertices, there will be separate distributions for each. However, their combination will not be a simple sum of electron curves, because each event is reconstructed as a single electron, with a single vertex, and the residuals are calculated relative to the fit position. The resulting distortion leads to measurable time differences in several important cases, discussed in the following section.

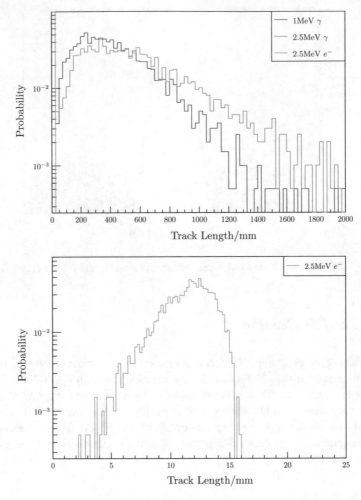

Fig. 3.11 Total track length for 2.5 MeV electrons, 2.5 MeV γ and 1.25 MeV γ in the SNO+ scintillator. Calculated as the length between the creation step and destruction step of the Geant4 track

Electrons

Figure 3.11 shows that, in the scintillator, 2.5 MeV electrons produce tracks of around 1 cm length. SNO+ is expected to reconstruct events 8.5 cm away from the true vertex position, on average, so information regarding the length and direction of the electron track is washed out by position resolution and the events appear 'point-like'. This is the reason electron reconstructed time residuals are close to the intrinsic time response of the scintillator (Fig. 2.12).

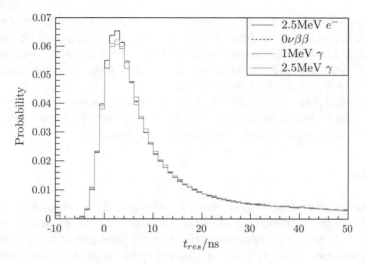

Fig. 3.12 Early reconstructed time residuals for electrons and γ, $r_{fit} < 3500$ mm

$0\nu\beta\beta$ and $2\nu\beta\beta$

Figure 3.12 shows that the reconstructed time residual spectra for $0\nu\beta\beta$ and 2.5 MeV electrons are identical, because timing resolution and vertex reconstruction wash out any difference between one electron track and two. $2\nu\beta\beta$ events will only differ from $0\nu\beta\beta$ events because their energy is, in general, different and time residuals run with energy.

γ

γ have no charge and therefore do not scintillate directly. However, they are detectable via secondaries produced in interactions with the scintillator. For a mostly Carbon target, the interaction cross-section of $\mathcal{O}(1\,\mathrm{MeV})$ γ is dominated by Compton scattering[4], which produces high energy scintillating electrons. A single γ will scatter 10s of times, producing electrons at each interaction site. The time delay between these scatters will be detectable, if the separation of the scatters in time and space is large enough.

The mean free path for such γ is easily estimated. Assuming that LAB has uniform chains of chemical formula $C_6H_5C_{13}H_{27}$, LAB is proportionately $\frac{57}{65}$ Carbon and $\frac{8}{65}$ Hydrogen by weight. The absorption lengths per unit density for Compton scattering on electrons bound in Hydrogen and Carbon are given by the PDG as $\lambda_H = 7\,\mathrm{g\,cm}^{-3}$, $\lambda_C = 19\,\mathrm{g\,cm}^{-3}$ at 1 MeV [4]. The total mean free path per unit density is then $\frac{1}{\lambda_{tot}} = \frac{w_H}{\lambda_H} + \frac{w_C}{\lambda_C}$ and, using $\rho_{LAB} = 0.863\,\mathrm{g\,cm}^{-3}$, the overall mean free path is $\lambda_{tot} \sim 18\,\mathrm{cm}$.

In reality, such γ scatter multiple times, depositing most of the energy in the first few scatters. Figure 3.11 shows the track lengths of 1 MeV and 2.5 MeV γ events;

the averages are 43 and 57 cm, respectively. LAB has a refractive index of ≈ 1.5 so 40 cm corresponds to a time delay of ~ 2 ns. This is comparable to both the PMT TTS (1 ns) and the fast scintillation time of LAB (8 ns) so the delay should be detectable: γ events will appear non-point-like. Indeed, Fig. 3.12 shows the time residuals are broader for the γ events than electron events.

Radioactive β decays are often followed by one or more nuclear de-excitation γ after ps time scales. In the ns scale SNO+, detector these will appear as a single event with both point-like and non-point-like components.

Positrons

Positions scintillate in an identical way to electrons, but their behaviour at the end of the track is different, where each position annihilates with a nearby electron, emitting γ. When a positron annihilates in vacuum there is a finite probability of forming an e^+e^- bound state called positronium. Combining two spin 1/2 particles allows for two possibilities: a spin 0 singlet state and a spin 1 triplet state. The former, para-positronium (p-Ps), decays to two γ within ps, while spin selection rules mean that the latter, ortho-positronium (o-Ps), decays after 138.6 ns to three γ. The degeneracy of the the triplet state means that o-Ps is formed in 3/4 of cases.

This picture is more complicated in dense matter; there, chemical reactions, magnetic effects and interactions with electrons cause o-Ps to p-Ps conversions, shortening the o-Ps lifetime and reducing the probability of its formation. These effects are medium dependent so the lifetime and formation fraction of o-Ps are also medium dependent. The formation fraction and o-Ps lifetime have been measured in LAB + H_2O + 0.3% Te solution to be (0.36 ± 0.009) and (2.69 ± 0.05) ns [5]. The SNO+ phase I loading will be higher, but the formation fraction and lifetime have been shown to be insensitive to the Te and H_2O concentrations in this range [5].

p-Ps decay is too quick to be distinguished from direct annihilation, but the 2.7 ns live-time of o-Ps is comparable to the detector resolution, and should be comfortably detectable in many cases.

Unfortunately, o-Ps simulation is not implemented in Geant4, so a custom physics process was added to the RAT simulation to handle positron annihilation. With 36% probability, it introduces a time delay between the end of the positron track and the creation of the annihilation γ. The size of the delay is sampled from an exponential probability distribution with characteristic time 2.7 ns. All events containing positrons in this work have been simulated with this modification except the production Monte Carlo used for $0\nu\beta\beta$ signal extraction in Chap. 8.

3.2.3 Hypothesis Testing for Event Classification

This final section addresses how best to classify an event as signal or background using its reconstructed time residuals and defines the techniques that will be applied in later chapters.

Typically, the time residuals are binned into a histogram with N_{bins} with bin contents $\{N^i\}$. The task is to condense that information in to a single discriminant $\mathcal{T}$ which can be used to distinguish $0\nu\beta\beta$ from background using a cut or a likelihood fit.

The simplest method places a hard cut on the time residuals and counts the fraction of hits above or below the corresponding critical bin i_{crit}. The fraction of hits before this critical time is f_{prompt}, a linear classifier of the form $\mathcal{T} = \sum_{i=0}^{N_{bins}} w^i N^i$ with weights defined by:

$$w^i = \begin{cases} 1 & i < i_{crit} \\ 0 & \text{otherwise} \end{cases} \tag{3.25}$$

f_{prompt} has no guaranteed optimal properties, but it is conceptually simple and robust.

The Neyman–Pearson lemma guides the choice for an optimal statistic. It states that, for any two *simple* hypotheses,[5] the likelihood-ratio test is uniformly most powerful [6]. In this context, a cut on the likelihood-ratio gives the highest $0\nu\beta\beta$ efficiency at any given background rejection. For event classification, the hypotheses are the different event types, signal or background $H = (S \text{ or } B)$, and the observables are the bin contents $\{N^i\}$, so the likelihood is the probability of observing a given histogram for an event of the type considered:

$$\mathcal{L}_H = \mathcal{L}(H|\{N^i\}) = P(\{N^i\}|H) \tag{3.26}$$

and the likelihood-ratio is:

$$\mathcal{T} = \frac{P(\{N^i\}|S)}{P(\{N^i\}|B)} \tag{3.27}$$

This work uses the convention that, for each likelihood-ratio considered, positive values indicate events with signal-like time residuals ($0\nu\beta\beta$) and negative values indicate background-like events. Unfortunately, the PDF in Eq. 3.26 has as many dimensions as there are bins in the histogram. To distinguish between the event classes in Fig. 3.12, this is comfortably 50D, which is intractable to estimate with Monte Carlo.

One remedy is to assume that each hit is independent, then the likelihood for each hit depends only on which time residual bin it falls into. If hit j falls into time residual bin b_j with probability $P(b_j|H)$:

$$\mathcal{L}_H = \Pi_{j=0}^{N_{hit}} P(b_j|H) = \Pi_{i=0}^{N_{bins}} P(b_i|H)^{N_i} \tag{3.28}$$

$P(b_j|H)$ is a 1D distribution which *can* be easily estimated from Monte Carlo, by calculating the average spectrum from a large number of events (e.g. Fig. 3.12). Combining the two hypotheses, the log-likelihood-ratio is:

---

[5]i.e. all of the parameters in the hypothesis are specified.

$$\mathcal{T} = \log \frac{\mathcal{L}_S}{\mathcal{L}_B} = \Sigma_{i=0}^{N_{bins}} N^i \log \frac{P(b_i|S)}{P(b_i|B)} \tag{3.29}$$

which is a linear classifier with weights $w^i = \log \frac{P(b_i|S)}{P(b_i|B)}$.

The accuracy of this assumption depends on the type of events tested. For point-like events like electrons, the SNO+ detector is insensitive to internal degrees of freedom such as track length, so the time residuals of each event look as though they are random draws from the same average distribution. In this case the independent hit assumption is likely sound.

For multi-vertex events the picture is less clear. A radioactive decay event emitting a β and two γ is a complex composite hypothesis with many unspecified parameters: the γ track lengths, the separation angles of the particles the etc. will vary between events. These internal degrees of freedom will create correlations. For example, a longer than average γ track length will create more late hits than average across several bins. Therefore, for these events, the late bins will be positively correlated with one another and anti-correlated with earlier times.

It can be useful to assume independence even when correlations are significant. In this context the expression in Eq. 3.29 is known as a 'Naive Bayes' classifier in the machine learning community. There it has been shown that the classifier performs well on correlated inputs, provided the correlations are similar between the event classes [7] e.g. a common reconstruction effect.

If correlations between the bins are significant, a simple extension is to use a Fisher discriminant. This is another linear classifier which explicitly considers correlations in the calculation of the weights:

$$\vec{w} = (\Sigma_S + \Sigma_B)^{-1} \cdot (\vec{\mu}_s - \vec{\mu}_b) \tag{3.30}$$

where $\vec{\mu}_H^i$ and Σ_H are the average normalised bin contents, and the covariance matrix between those bins for hypothesis H, respectively. The Fisher discriminant maximises the separation between the average classifications of the classes, divided by the variance of the classification within the classes:

$$\frac{\bar{T}_S - \bar{T}_B}{\text{Var}(T_S) + \text{Var}(T_B)} \tag{3.31}$$

If the hypotheses are Gaussian with common covariance this is equivalent to the full correlated likelihood-ratio [6].

The Gatti parameter is another linear classifier used by the Borexino collaboration, with weights are defined by:

$$w^i = \frac{\mu_S^i - \mu_B^i}{\mu_S^i + \mu_B^i} \tag{3.32}$$

On close inspection, the Gatti parameter, is just a special case of the Fisher discriminant in which the the hits are independent. Then, the content of each bin is Poisson distributed with mean μ^i and $\Sigma = \text{diag}(\mu^0, \mu^1, \mu^2..)$. Substituting these conditions into Eq. 3.30 recovers Eq. 3.32. The Gatti parameter is often claimed optimal (e.g. [8]), but it is only valid when the hits are independent and, even in those circumstances, one is able to use the likelihood-ratio provided in Eq. 3.29, which has guaranteed optimal properties.

If the data are highly correlated and not linearly separable, better results may come from machine learning algorithms, because many have the advantage of being able to learn non-linear classification boundaries. If the data are not linearly separable, such algorithms can drastically out perform those described above. Conversely, if the data are close to linearly separable, such algorithms can add much to complexity and little to efficacy.

References

1. Coulter I (2013) Modelling and reconstruction of events in SNO+ related to future searches for lepton and baryon number violation. Ph.D. thesis
2. Jones P (2011) Background rejection for the neutrinoless double beta decay experiment SNO+. Ph.D. thesis
3. Press William H, Teukolsky Saul A, Vetterling William T, Flannery Brian P (2007) Numerical recipes 3rd edition: the art of scientific computing, 3rd edn. Cambridge University Press, New York
4. Olive KA et al (Particle Data Group) (2014) Review of particle physics. Chin Phys C38:090001. https://doi.org/10.1088/1674-1137/38/9/090001
5. Franco D, Consolati G, Trezzi D (2011) Positronium signature in organic liquid scintillators for neutrino experiments. Phys Rev C-Nucl Phys 83(1). https://doi.org/10.1103/PhysRevC.83.015504, arXiv:1011.5736
6. Cowan G (1998) Statistical data analysis. Oxford Science Publications, Clarendon Press, Oxford. https://books.google.co.uk/books?id=ff8ZyW0nlJAC
7. Zhang H (2004) The optimality of naive bayes, vol 2
8. Back HO et al (2008) Pulse-shape discrimination with the counting test facility. Nucl Instrum Methods Phys Res A 584:98–113. https://doi.org/10.1016/j.nima.2007.09.036. arXiv:0705.0239

Chapter 4
Backgrounds

At current maximum allowed ^{130}Te $0\nu\beta\beta$ half-life of $T_{1/2} = 1.5 \cdot 10^{25}$yr, SNO+ would expect 278 signal counts per year at the $2\nu\beta\beta$ end point, 2.5 MeV, smeared by an energy resolution of 82 keV. Unfortunately, there are many significant sources of background in this region. Claiming a $0\nu\beta\beta$ signal, or setting a stringent limit on its rate, requires that these backgrounds are, first, kept to a minimum and, second, that they are well constrained to mitigate systematic error.

This chapter reviews each of the expected major backgrounds: their origin; their expected rates; and the handles that may be used to reject or constrain them. The final section of this chapter defines the SNO+ signal window, the projected background budget inside it and the sensitivity of SNO+ to $0\nu\beta\beta$ using a simple counting analysis.

4.1 Solar Neutrinos

During the tellurium phase, SNO+ will detect neutrinos which are produced in the sun and free stream to earth before interacting with the scintillator target. These neutrinos can be a $0\nu\beta\beta$ background. Of particular concern are $\nu-e$ elastic scattering events which produce single electrons. In SNO+, these events are indistinguishable from $0\nu\beta\beta$ using timing or by any other means.[1] Figure 4.1 shows the neutrino fluxes at the surface of the earth; only ^{8}B spectrum neutrinos contribute significantly to the signal window. The rate of these events will be well constrained by SNO measurements of the ^{8}B flux to within 4% [3], but additional uncertainty arises from the shape of the ^{8}B spectrum and the dependence of the ν_e survival probability with energy [4].

---

[1] There are differences in the Cherenkov signals between the two, but the SNO+ detector is insensitive to these [1, 2].

© Springer Nature Switzerland AG 2019
J. Dunger, *Event Classification in Liquid Scintillator Using PMT Hit Patterns*,
Springer Theses, https://doi.org/10.1007/978-3-030-31616-7_4

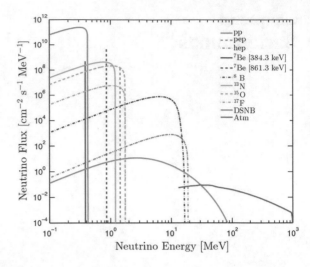

Fig. 4.1 Neutrino fluxes at the earth's surface [5]. DSNB = diffuse supernova background, Atm = atmospheric neutrinos

4.2 Double Beta Decay

Another major background comes from ordinary $2\nu\beta\beta$ decays: finite energy resolution pushes events from the steeply falling $2\nu\beta\beta$ tail into the signal window. $0\nu\beta\beta$ and $2\nu\beta\beta$ differ from one another only in the amount of visible energy released, so this background cannot be reduced without improved energy resolution.

The half-life of the decay is already well constrained by CUORE-0 to be $8.2 \pm 0.2(\text{stat}) \pm 0.6(\text{sys}) \times 10^{20}\text{yr}$ [6] and it is likely that the CUORE experiment will publish an improved measurement on the half-live and energy spectrum in time for use in SNO+ analysis [6]. This half-life translates to one $2\nu\beta\beta$ event every 6s in SNO+ phase I. The CUORE measurements could be used as a constraint on the $0\nu\beta\beta$ rate, provided the number of ^{130}Te nuclei in the target can be accurately determined.

In addition, the $2\nu\beta\beta$ background should dominate the events collected in the region 1.5–2.5 MeV. There, its steeply falling energy spectrum will be powerful for extracting the $2\nu\beta\beta$ normalisation, provided energy response uncertainties can be kept under control [4].

4.3 Natural Internal Radioactivity

There will also be background counts from natural radioactive isotopes inside the scintillator cocktail. The most important isotopes are the daughters of ^{238}U and ^{232}Th, for which the decay chains are shown in Fig. 4.2. In this work, the chains are assumed to be in equilibrium.[2]

---

[2]I.e. they have had time to reach steady state, without injection of isotope mid-chain. Equilibrium is often broken between ^{238}U and ^{226}Ra, but this is unimportant for $0\nu\beta\beta$ searches, for which all of the significant backgrounds are below ^{226}Ra in the decay chain.

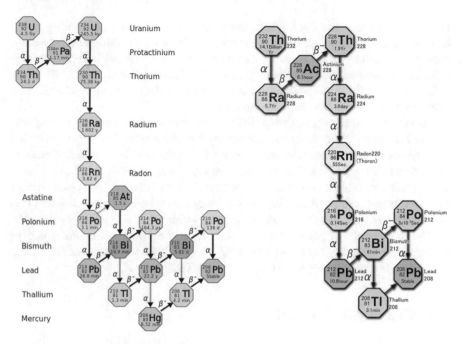

Fig. 4.2 The uranium and thorium decay chains [7, 8]

The most problematic isotopes are ^{214}Bi and ^{212}Bi. Both are $\beta\gamma$ decays with endpoint values of 3.27 MeV/2.25 MeV, respectively. $^{212/214}$Bi decays are followed by $^{212/214}$Po α decays after characteristic times of 0.3/164 µs. These decays have Q values of 8.95/7.7 MeV which are typically quenched down to visible energies of 0.9 MeV and 0.75 MeV, respectively.[3] The Bi and Po events can fall into a single trigger window or two, depending on the delay between the two; both cases are problematic. If the decays fall in separate windows, the Bi decay can fall into the signal region, whereas if the decays fall into the same trigger window, their combined energy can do the same.

Less important, but significant, is ^{210}Tl, a β decay isotope. It has $Q = 5.5$ MeV, leading to overlap of its reconstructed energy spectrum with the region of interest.

The primary strategy for removing these backgrounds is purification of the scintillator. SNO+ targets LAB impurity levels equal to those measured by Borexino [9–11] and aims to purify the Diol to the same level as the external water [9]. Tables A.2 and A.3 in Appendix A show the each of the target concentrations of these contaminants and their resulting yearly counts across the detector. Even after purification, significant suppression is required: SNO+ would expect 4.73×10^4 and 682 counts per year of ^{214}BiPo and ^{212}BiPo, respectively, in the region $Q_{\beta\beta} \pm \sigma_E$.

Fortunately, the delayed coincidence between the Bi and Po decays can be used to tag these events. Two trigger events will be tagged using the proximity of the

---

[3] According to RAT6.1.6 simulation.

two events in space and time, whilst events that pile up into a single trigger will be rejected using the double pulse structure of their hit time residuals [12]. Overall, these techniques allow for rejection factors of $\approx 5 \times 10^5$ with $<1\%$ $0\nu\beta\beta$ sacrifice [13]. The eliminated events will also provide a useful tagged sample of ^{214}Bi events that may be used for calibrations and, if equipped with evidence of equilibrium, estimations of other U/Th chain contaminants.

Simulation studies suggest that the BiPo events that pass the coincidence cuts come in two categories. First, those two trigger events where a high energy γ produced in ^{214}Bi decay travels several meters away from the initial decay before Compton scattering. In this case the Bi and Po vertices are too far apart for them to be associated without also rejecting an unacceptable number of $0\nu\beta\beta$ events. The second type is those one trigger events where the β and α decay are separated by less than 5 ns [12, 13]. These will look like single vertex events to within the resolution of the detector.

The final handle is pulse shape discrimination. The long tail of Po α decays distinguishes them from electron like $0\nu\beta\beta$ and the multi-site nature of the Bi and Tl $\beta\gamma$ decays distinguishes them from point-like $0\nu\beta\beta$ events.

4.4 External Radioactivity

Other significant sources of radioactivity lie outside of the scintillator volume. Such 'external' backgrounds originate in the external UPW, the volume of the AV, dust on the inner/outer AV surfaces, the PMTs and the rope systems. The isotopes of most concern are ^{208}Tl and ^{214}Bi, which produce low energy particles outside of the AV and >1 MeV γ that can penetrate into the detector centre.

Figure 4.3 shows the expected external background contribution of those events that reconstruct inside the AV, as a function of energy. ^{208}Tl decay always produces a particularly dangerous 2.6 MeV γ, which leads to a strong peak close to $Q_{\beta\beta}$, while ^{214}Bi can produce a range of γ, giving it a broad energy spectrum which overlaps with the signal region in parts. The rate of both ^{208}Tl and ^{214}Bi decays dwarfs any realistic signal.

Fig. 4.3 Reconstructed energy spectra for external ^{208}Tl and ^{214}Bi decay inside the AV, normalised to the expected rate in that region

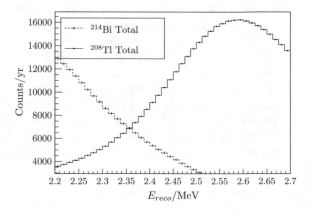

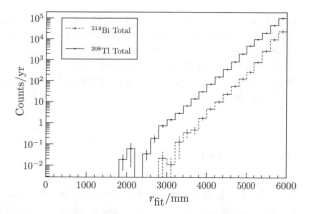

Fig. 4.4 Yearly external background rates vs event radius in the AV, normalised to the expected rate in that region

The primary strategy for these events is to fiducialise away the outer parts of the detector, where these decays originate. However, they can still present a background to $0\nu\beta\beta$ because the Compton scattering length of the 2.6 MeV γ is 10 cm in water and 26 cm in LAB. This means that a very small fraction of these γ travel into the detector centre, producing an event with apparent energy close to 2.6 MeV. The expected yearly rates for these external backgrounds are shown as a function of fitted event radius, r_{fit}, in Fig. 4.4. Within the detector centre, where the external backgrounds are lowest, the biggest contributors are ^{208}Tl decays from the external water, hold down ropes, acrylic vessel and the PMTs.

Those decays that do penetrate into the detector centre can be distinguished from real $0\nu\beta\beta$ using light produced by particles outside the AV and using the time smearing caused by the multiple scattering of the γ inside the scintillator volume. These handles are explored in more detail in Chap. 6.

The external event rates assumed in this work are shown in Table A.4 of Appendix A. Constraints on these normalisations will come from measurement in the water and scintillator phases and the high radius side-band during the tellurium phase [4].

4.5 Leaching

During the years between draining the SNO experiment in 2006 and refilling in 2016, the inside of the AV was exposed to airborne radon which plated onto the surface, to a depth of several hundred nm. In days, ^{222}Rn decays to ^{210}Pb which has a half-life of 22.2 years. The result is a layer of ^{210}Pb, several hundred nm thick, which steadily feeds the detector with its daughters: ^{210}Po and ^{210}Bi. Measurements of the AV suggest activities of 1.18 kBq for both [9].

The quenched energy deposited in these decays is too low to form a direct background to $0\nu\beta\beta$, unless the rate is much higher than expected and the decays pile up significantly. However, a large rate can contribute to signal sacrifice in coincidence cuts, the α produced can create free neutron backgrounds (Sect. 4.6) and low energy pile-ups can complicate side bands used to constrain other backgrounds.

Similarly, ^{222}Rn from the external water and PMT surfaces can diffuse through the AV acrylic into the scintillator. This will produce additional background from ^{210}Pb and its daughters [14].

4.6 α–n Decays

$\alpha-n$ backgrounds occur when α created in radioactive decay are absorbed by a nucleus, causing the nucleus to emit a neutron, a daughter nucleus and sometimes associated particles.

These events produce several signals: first, the α produces scintillation light; second the neutron scatters into protons causing them to scintillate; third, once thermalised, the neutron captures on another nucleus, possibly causing it to γ decay. Most common is capture on hydrogen to produce a 2.2 MeV γ. Both the prompt signal (from the α and protons) and the neutron capture can reconstruct close to the $0\nu\beta\beta$ signal window, forming a background.

The dominant source of α in the SNO+ detector will be ^{210}Po nuclei on the AV surface and leached into the scintillator volume. The exact rate of these decays depends acutely on the leaching model but 6 possible contributions have been identified, shown in Table A.5 of Appendix A. At current leaching assumptions, the event rate is greatest for AV α decay but decays in the scintillator form the largest contribution to the signal window inside the detector centre [15].

The neutron capture signal is delayed relative to the prompt signal by a thermalisation time of 220 μs and an average path length of 49 cm [16]. $\alpha-n$ backgrounds can be tagged with high efficiency using this delayed coincidence (a cut is applied in Chap. 8).

4.7 Cosmogenics Isotopes

Cosmogenic isotopes are not naturally abundant but can be created by the spallation of cosmic rays on natural materials at the earth's surface [17]. The decay of these isotopes can form a background to $0\nu\beta\beta$. Heavy tellurium[4] is of particular concern as a target for producing long lived, radioactive isotopes with high-Q values. The SNO+ tellurium will be exposed to a significant cosmic ray flux at the surface, increasing the contamination of these cosmogenic isotopes until the Te is taken underground to SNOLAB.

---

[4]Relative atomic mass of 127.6 u.

These isotopes require a more in-depth treatment than the backgrounds discussed so far, because they are expected to be removed almost completely in underground tellurium purification. What makes many of them dangerous is their ability to mimic a $0\nu\beta\beta$ signal if the purification is incomplete. For this reason, the next section identifies the most problematic cosmogenic isotopes in terms of their degeneracy with a $0\nu\beta\beta$ signal, rather than their expected rates.

4.7.1 Problem Isotopes

Lozza and Petzoldta calculated the possible cosmogenic isotopes produced by spallation of cosmic rays on a natural tellurium target at the earth's surface [17]. Table 4.1 shows the 18 candidates they identified that satisfy $Q > 2\,\mathrm{MeV}$, $Z < 131$ and $T_{\frac{1}{2}} > 20\mathrm{d}$ (shorter half-lives have been included if they are fed by a longer lived parent).

Table 4.1 Cosmogenic $0\nu\beta\beta$ background candidates produced on Tellurium [17]. Each isotope has $Q > 2\,\mathrm{MeV}$, $Z > 131$ and $T_{\frac{1}{2}} > 20\mathrm{d}$. (m) denotes meta-stable states

Isotope	$T_{\frac{1}{2}}$/days	Q/MeV	Decay mode (BR)
^{22}Na	950.6	2.84	EC/β^+
^{26}Al	2.62×10^8	4.00	β^+
^{42}K (direct and daughter of ^{42}Ar)	0.51 (1.2×10^4)	3.53	β^-
^{44}Sc (direct and daughter of ^{44}Ti)	0.17 (2.16×10^4)	3.65	EC/β^+
^{46}Sc	83.79	2.37	β^-
^{56}Co	77.2	4.57	EC/β^+
^{58}Co	70.9	2.31	EC/β^+
^{60}Co (direct and daughter of ^{60}Fe)	1925.27 (5.48×10^8)	2.82	β^-
^{68}Ga (direct and daughter of ^{68}Ge)	4.70×10^{-2} (2.71)	2.92	EC/β^+
^{82}Rb (daughter of ^{82}Sr)	8.75×10^{-4} (25.35)	4.40	EC/β^+
^{84}Rb	32.8	2.69	EC/β^+ (96.1)
^{88}Y (direct and daughter of ^{88}Zr)	106.63 (83.4)	3.62	EC/β^+
^{90}Y (direct and daughter of ^{90}Sr)	2.67 (1.05×10^4)	2.28	β^-
^{102}Rh (direct and daughter of ^{102}Rh(m))	207.3	2.32	EC/β^+ (78)
^{102}Rh(m)	1366.77	2.46	EC (99.77)
^{106}Rh (direct and daughter of ^{106}Ru)	3.47×10^{-4} (371.8)	3.54	β^-
^{110}Ag(m)	249.83	3.01	β^- (98.67)
^{110}Ag (daughter of ^{110}Ag(m))	2.85×10^{-4}	2.89	β^- (99.70)
^{124}Sb	60.2	2.90	β^-
^{126}Sb(m) (direct and daughter of ^{126}Sn)	0.01 (8.40×10^8)	3.69	β^- (86)
^{126}Sb (direct and daughter of ^{126}Sb(m))	12.35 (0.01)	3.67	β^-

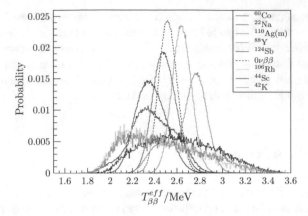

Fig. 4.5 Reconstructed energy distributions for cosmogenic $0\nu\beta\beta$ backgrounds and $0\nu\beta\beta$ signal events

Very long lived isotopes produce slow event rates for a given normalisation and very short lived isotopes decay away during a 'cool down' period. More worrisome are the $\mathcal{O}(1\mathrm{yr})$ half-lives that are comparable with the likely run-time of the experiment. In this regard the problematic isotopes are ^{110,110m}Ag, ^{56,58,60}Co, ^{22}Na, ^{102,102m,106}Rh, ^{124,126}Sb, ^{106}Rh, ^{44,46}Sc, ^{42}K and ^{88,90}Y.

Cosmogenic backgrounds only cause difficulty if they have similar energy spectra to the expected signal and fortunately, few of them overlap significantly with $0\nu\beta\beta$. (^{56,58}Co, ^{90}Y) have negligible probability of reconstructing in the $0\nu\beta\beta$ signal region.[5] $^{102,102(m)}$Rh, ^{126}Sb, ^{46}Sc do contribute to the signal region but each have a peak elsewhere that contains orders of magnitude more probability; if these isotopes did contribute a background to the signal region, the peaks outside the window would be easily identified. ^{110}Ag may also be discounted because its parent, 110Ag(m), only produces it in 1.33% of decays, so the latter could always be measured from the former.

The remaining isotopes ^{44}Sc, ^{42}K, ^{60}Co, ^{22}Na, ^{106}Rh, ^{110}Ag(m) and ^{88}Y have both significant energy overlap and $\mathcal{O}(1\mathrm{yr})$ half-lives. These are as the 'problem isotopes' for study in Chap. 5. Figure 4.5 shows their energy spectra alongside the $0\nu\beta\beta$ signal and Fig. 4.6 shows their cool down over several years.

Lozza and Petzoldta calculated the production rate of each of these isotopes [17], the results are shown in Table A.1 of Appendix A. They find that, without purification, these isotopes would contribute thousands of events per tonne of Te in the experiment's first year, after 1 yr exposure at the surface. All considered, the most dangerous isotopes are ^{60}Co, ^{88}Y and ^{22}Na. They are formed in large numbers during surface exposure, they have half-lives of several years and their energy distributions are similar in shape and scale to the $0\nu\beta\beta$ signal.

---

[5]According to RAT 6.1.6 simulation.

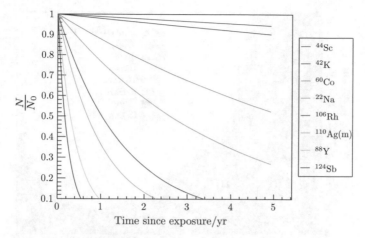

Fig. 4.6 Cool down of problem cosmogenic isotopes. In cases where very short lived isotopes are fed by long lived parents, the parent's half-live is shown

4.7.2 Mitigation

Using underground Te purification, SNO+ aims to reduce the yearly cosmogenic event rate from several thousand per tonne of Te, to a *total* event rate of 0.19 cts/yr [9]. These event rates are insignificant, more difficult is constraining the rates of these isotopes in-situ. There are three available handles. First, their characteristic decay times can be used to distinguish them from $0\nu\beta\beta$: if there were a significant contamination of one or several of these isotopes the event rate should cool down according to Fig. 4.6, whereas a $0\nu\beta\beta$ signal rate would change only with additional loading. Second, the contamination can be constrained ex-situ using measurements of the pre-purification tellurium activity and the purification efficiency can be estimated using spike tests. Finally, many of these decays emit γ or e^+. This gives those decays a characteristic multi-site signature that differs from point-like $0\nu\beta\beta$ decays. Discrimination based on this principle is explored in detail in Chap. 5.

4.8 Counting Analysis

The simplest SNO+ $0\nu\beta\beta$ analysis defines a signal region in reconstructed energy and event radius and uses it to perform a counting experiment. There are only a few expected external counts per year inside the central 3.5 m. This is the SNO+ fiducial volume, chosen to maximise the detector sensitivity: inside this region, other backgrounds, particularly ^{8}B ν ES events, dominate over the externals [18]. In addition, a region of interest in energy is defined around $Q_{\beta\beta}$, to isolate the $0\nu\beta\beta$ signal from most backgrounds; for this, an asymmetric window is chosen in order to offset the

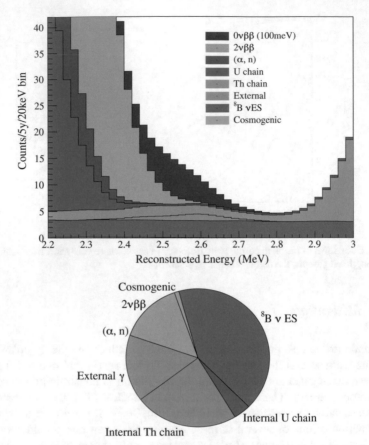

Fig. 4.7 $0\nu\beta\beta$ background budget and energy spectrum near $Q_{\beta\beta}$ inside the fiducial volume [15]

effect of the steeply falling $2\nu\beta\beta$ spectrum below $Q_{\beta\beta}$:

$$Q_{\beta\beta} - 0.5\sigma_E \rightarrow Q_{\beta\beta} + 1.5\sigma_E = (2.49 - 2.65) \text{ MeV} \qquad (4.1)$$

Figure 4.7 shows the expected energy spectrum near $Q_{\beta\beta}$ after 5 years live-time, inside the fiducial volume and assuming a $0\nu\beta\beta$ signal equivalent to $m_{\beta\beta} = 100$ meV. The signal presents as an excess of events on top of the falling $2\nu\beta\beta$ edge. Inside the signal window, the background is dominated by ^{8}B ν ES events, with sub-dominant contributions from radioactive backgrounds.

A statistics only counting analysis using this signal window gives an expected limit on the $0\nu\beta\beta$ half-life of $T_{1/2} > 1.9 \times 10^{26}$ years at 90% confidence, after a 5 year live-time [15]. Figure 4.8 shows the best published half-life [19] and $m_{\beta\beta}$ limits as of August 2017[6] [19] alongside this projected sensitivity. If achieved, the SNO+ limit will be world leading.

---

[6]CUORE have since improved their limit to 1.5×10^{25} [20].

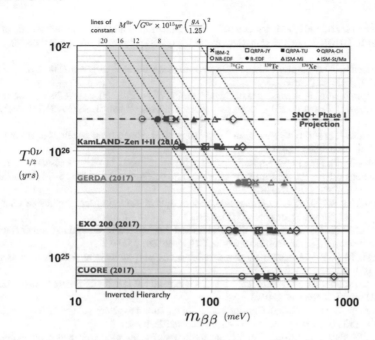

Fig. 4.8 Comparison between existing $0\nu\beta\beta$ limits [19]. Diagonals indicate the (unknown) constant of proportionality between $T_{1/2}^{0\nu}$ and $m_{\beta\beta}$, which is determined by nuclear physics. Each of the points shows a theoretical prediction of this constant for the $0\nu\beta\beta$ nucleus used in each experiment

Cosmogenic decays are a negligible contribution to the background budget. However, the energy spectra of these backgrounds make them degenerate with the $0\nu\beta\beta$ signal. This severely limits the experiment's discovery potential: without proper constraint, the excess in Fig. 4.7 could just as easily be ^{60}Co. Indeed, what makes these isotopes particularly tricky is that they are introduced with the tellurium, so their rate will scale linearly with the Te-loading, in the same way as a $0\nu\beta\beta$ signal would. Thus the need for an in-situ constraint of these decays, the focus of Chap. 5.

References

1. Beier G (2016) Study of direction reconstruction in scintillator. SNO+-docDB 3503-v4
2. Mottram M (2015) Cherenkov reconstructieon in scintillator at 2.5 MeV. SNO+-docDB 3116-v2
3. Aharmim B et al (SNO) (2013) Combined analysis of all three phases of solar neutrino data from the Sudbury neutrino observatory. Phys Rev C88:025501. https://doi.org/10.1103/PhysRevC.88.025501, arXiv:1109.0763
4. Mastbaum A (2015) Systematics and constraints in the double beta counting analysis. SNO+-docDB 3000-v1 (2015)
5. O'Hare CAJ (2016) Dark matter astrophysical uncertainties and the neutrino floor. arXiv:1604.03858v1

6. Alduino C et al (2017) Measurement of the two-neutrino double-beta decay half-life of 130 Te with the CUORE-0 experiment. Eur Phys J C. https://doi.org/10.1140/epjc/s10052-016-4498-6

7. Wikipedia, The uranium chain. https://commons.wikimedia.org/wiki/File:Decay_chain(4n%2B2,_Uranium_series).svg

8. Wikipedia. https://en.wikipedia.org/wiki/Thorium#/media/File:Decay_Chain_Thorium.svg

9. O'Keeffe H, Lozza V (2017) Expected radioactive backgrounds in SNO+. SNO+-docDB 507-v35

10. Arpesella C et al (2008) New results on solar neutrino fluxes from 192 days of Borexino data, pp 1–6. arXiv:0805.3843v2

11. Alimonti et al G (2009) Nuclear instruments and methods in physics research a the liquid handling systems for the Borexino solar neutrino detector, vol 609, pp 58–78. https://doi.org/10.1016/j.nima.2009.07.028

12. Majumdar K (2015) On the measurement of optical scattering and studies of background rejection in the SNO+ detector, Ph.D. thesis

13. Wilson JR (2016) Summary of the uranium and thorium chain background considerations for a low energy ^{8}B solar neutrino analysis. SNO+-docDB 3531-v1

14. Wojcik M (1991) Measurement of radon diffusion and solubility constants in membranes, vol 61, pp 8–11

15. Kaptanoglu T (2016) Te Diol 0.5% loading approved BB sensitivity plots and documentation. SNO+-docDB 3689-v2 (2016)

16. Andringa S et al (2016) Current status and future prospects of the SNO+ experiment. arXiv:1508.05759v3 [physics. ins-det]. Accessed 28 Jan 2016

17. Lozza V, Petzoldt J (2015) Cosmogenic activation of a natural tellurium target. Astropart Phys 61:62–71. https://doi.org/10.1016/j.astropartphys.2014.06.008. arXiv:1411.5947

18. Coulter I (2013) Modelling and reconstruction of events in SNO+ related to future searches for lepton and baryon number violation, Ph.D. thesis

19. Biller S (2017) 0nbb sensitivity comparison plot. SNO+-docDB 4769

20. Alduino C et al (CUORE) (2017) First Results from CUORE: a search for lepton number violation via $0\nu\beta\beta$ Decay of ^{130}Te. arXiv:1710.07988

Chapter 5
Pulse Shape Discrimination for Internal Backgrounds

This chapter applies the reconstructed time residual technique outlined in Sect. 3.2 to discriminating between $0\nu\beta\beta$ and internal $\beta^{\pm}\gamma$ radioactive backgrounds. The first section deals with separating $0\nu\beta\beta$ events from the problem cosmogenic isotopes identified in Sect. 4.7. Other than decay half-life, this is the only method for in-situ constraint of these isotopes, which are strongly degenerate with the $0\nu\beta\beta$ signal. The second section deals with the possibility of using the same technique to further reduce the background count from uranium and thorium chain contaminants in the scintillator. The last section explores the method's robustness and the expected signals that could be used to calibrate it.

Each of the data sets in this chapter was generated uniformly in the AV volume and across the full energy range of the decay, but cuts were applied to select only events that reconstructed as valid, within an energy signal window of $2.438 < E/\text{MeV} < 2.602$[1] and within the fiducial volume of $r < 3.5$ m. In each case, the events used to tune the classifiers are completely independent from the events used to test them.

5.1 Cosmogenics

There are many cosmogenic isotopes which will be produced in cosmic ray spallation on the tellurium while it is at surface. Chapter 4 identified seven of the isotopes that can be produced on Te as particularly dangerous for a $0\nu\beta\beta$ search; these were: ^{44}Sc, ^{42}K, ^{60}Co, ^{22}Na, ^{106}Rh, ^{110}Ag(m) and ^{88}Y. These isotopes are expected to be produced in large numbers at the surface, they have half-lives comparable to the length of the experiment and their energy spectra strongly overlap with that of the $0\nu\beta\beta$ signal.

---

[1]This is $\pm$ one energy resolution from $Q_{\beta\beta}$.

© Springer Nature Switzerland AG 2019
J. Dunger, *Event Classification in Liquid Scintillator Using PMT Hit Patterns*,
Springer Theses, https://doi.org/10.1007/978-3-030-31616-7_5

5.1.1   Event Topology

The seven cosmogenic problem isotopes are either β^- or β^+/EC emitters, but they are better grouped by the most energetic particle of their most common decay mode. The first group is the β^- emitters: ^{42}K and ^{106}Rh emit no γ 82% and 76% of the time, respectively. Thus, the dominant decay mode of these decays will be difficult to separate from the $0\nu\beta\beta$ signal with timing, because both are electron-like. However, there are sub-dominant decays with γ: 8% of ^{42}K decays emit a 1.1 MeV γ and the sub-dominant modes of ^{106}Rh almost always produce a single 0.5 MeV γ.

The γ group is ^{60}Co, ^{124}Sb, ^{110}Ag(m) and ^{88}Y. The first three have dominant mode β end points of 0.3 MeV, 0.1/0.4 MeV and 0.7 MeV respectively, with the rest of the energy deposited by $\mathcal{O}(1\,\text{MeV})$ γ. ^{88}Y decay is dominated by electron capture decay with the emission of two γ of $\approx$1 MeV. These non-point-like events should have broader time profiles than point-like $0\nu\beta\beta$ events.

The final, $\beta^+\gamma$, group is ^{22}Na, and ^{44}Sc. The dominant mode for these decays is β^+ decay and they come with associated γ of 1.2 MeV and 1.3 MeV, respectively. These events will share the broadening of the γ group but there should also be a distinct sub-population of decays with very late light produced by o-Ps formation in 1/3 of cases.

5.1.2   Time Residuals

Figure 5.1 shows the reconstructed time residuals of these candidate isotopes and the $0\nu\beta\beta$ signal. The only feature in the spectrum sharp enough to be noticeably distorted by a $\mathcal{O}(1\,\text{ns})$ delay from Compton scattering or o-Ps formation is the central prompt peak.

Figures 5.2, 5.3 and 5.4 show these spectra grouped into β^-, γ and β^+ dominated decays. Within each group, the cosmogenic decays have near identical spectra, reflecting their common topology, but, in each case, the non-point-like cosmogenic timing spectra are broader than point-like $0\nu\beta\beta$. The decays in the β group have only a very modest difference from $0\nu\beta\beta$ events, which is driven by the sub-dominant γ decay modes. ^{42}K is noticeably broader than ^{106}Rh because its sub-dominant mode emits a higher energy γ. The γ group shows the expected broadening relative to $0\nu\beta\beta$ events, though the difference is at most 10% and confined to around 20 bins. Finally, the $\beta^+\gamma$ group shows the largest difference with $0\nu\beta\beta$. These decays produce the smallest prompt peak and the most hits on its falling edge.

It is important to note that, for each of the groups, the cosmogenics appear to produce more hits on the *rising* edge as well as the falling edge. The cosmogenics in Figs. 5.2, 5.3 and 5.4 all have more hits for $t_{res} < 0\,\text{ns}$ than $0\nu\beta\beta$ events. This is a reconstruction effect: Compton scattering and the o-Ps lifetime do push prompt hits in cosmogenic decays to later times than $0\nu\beta\beta$ events, but, the picture is complicated by vertex position reconstruction. `PositionTimeLikelihood` reconstructs these

Fig. 5.1 Time residual PDFs for problematic cosmogenic decays and $0\nu\beta\beta$. Spectra shown are the normalised means of 40000 events in the target volume, reconstructing in $r_{reco} < 350$ cm. The first bin is an underflow. The distribution extends to an overflow bin at zions

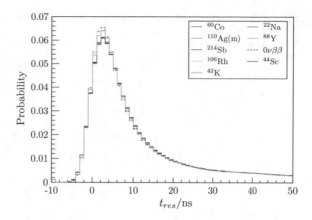

Fig. 5.2 Early time residuals for β^- dominated decays. Valid fits only, $r < 3.5$ m, 2.43 MeV $< E < 2.60$ MeV

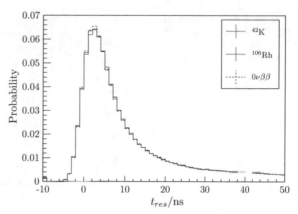

events by comparing them with the expected electron PDF. The algorithm selects a vertex that aligns the prompt peak of the event with the prompt peak of the PDF, making the earliest hits appear as though they occurred before the event.

The following three sections use increasingly complex discriminants to separate $0\nu\beta\beta$ from cosmogenics using the distributions in Figs. 5.2, 5.3 and 5.4.

5.1.3 f_{prompt}

The simplest discriminant for exploiting the differences between the time residual spectra is f_{prompt}, the fraction of hits before a critical time:

$$f_{prompt} = \frac{N_{hit}(t_{res} < t_{crit})}{N_{hit}} \tag{5.1}$$

Fig. 5.3 Early time residuals
for γ dominated decays.
Valid fits only, $r < 3.5$ m,
2.43 MeV $< E < 2.60$ MeV

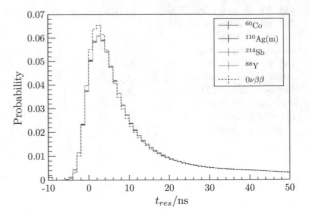

Fig. 5.4 Early time residuals
for $\beta^{+}\gamma$ dominated decays.
Valid fits only, $r < 3.5$ m,
2.43 MeV $< E < 2.60$ MeV

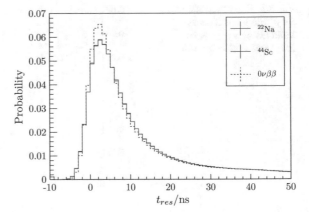

Figure 5.5 shows the relative expected significance for a f_{prompt} cut used to distinguish between $0\nu\beta\beta$ and cosmogenic backgrounds. It suggests that a critical time
of 4 ns is optimal for the f_{prompt} classifier.[2] The plot also suggests that a cut at -3 ns
could work equally well. However, the expected number of hits with $t_{res} < -3$ ns for
$0\nu\beta\beta$ events is only around 10, where the Gaussian approximation used to generate
the figure breaks down.

Figures 5.6, 5.7 and 5.8 show the f_{prompt} distributions for the three decay classes. As
expected, there is essentially no power to separate the β^{-} class, except for potentially
a tail from the sub-dominant ^{42}K γ decay. The $\beta^{+}\gamma$ and γ classes show separation
of around $1/2$ of one RMS width and the $\beta^{+}\gamma$ events show a clear low f_{prompt} tail,
caused by long-lived o-Ps states.

---

[2]Note f_{prompt} and C differ in that the former is calculated event-by-event and the latter is calculated
from the *average* spectrum.

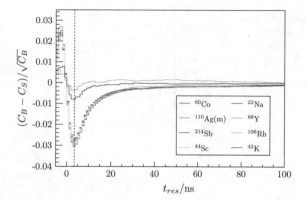

Fig. 5.5 Optimisation of the f_{prompt} cut for separating $0\nu\beta\beta$ from cosmogenic backgrounds. C_S, C_B are the cumulative probability distributions for the time residuals produced in $0\nu\beta\beta$ and cosmogenic events. The expression on the y-axis is related to the expected statistical separation power by a factor of $\sqrt{N_{hit}}$, provided that the number of early hits is Gaussian distributed

Fig. 5.6 f_{prompt} distributions for β^- dominated decays. Valid fits only, $r < 3.5$ m, 2.43 MeV $< E < 2.60$ MeV

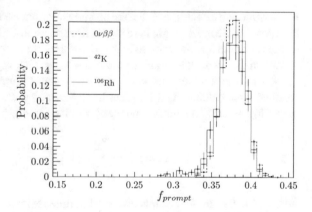

Fig. 5.7 f_{prompt} distributions for γ dominated decays. Valid fits only, $r < 3.5$ m, 2.43 MeV $< E < 2.60$ MeV

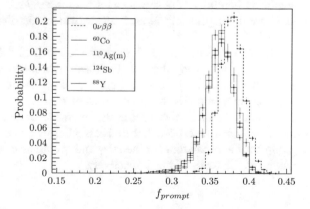

Fig. 5.8 f_{prompt} distributions
for $\beta^+\gamma$ decays. Valid fits
only, $r < 3.5$ m, 2.43 MeV
$< E < 2.60$ MeV

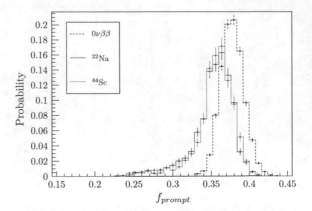

5.1.4 $\Delta \log \mathcal{L}$

Reconstruction effects mean that the cosmogenics have additional early hits as well as additional late hits. A single hard cut on time residuals, like f_{prompt} does not make use of the former effect. In order to make use of the excess of early hits for cosmogenics shown in Fig. 5.5, one could separate the $t_{res} < 5$ ns region into two regions, one $t_{res} < 0$ ns and another 0 ns $< t_{res} < 5$ ns. A better approach is to use all 210 time residual bins, calculating a log-likelihood for each event under the cosmogenic and $0\nu\beta\beta$ hypotheses. Assuming independent hits, the log-likelihood ratio is:

$$\Delta \log \mathcal{L} = \sum_{j=0}^{N_{bin}} N_j \log \left(\frac{P(b_j|0\nu)}{P(b_j|C)} \right) \qquad (5.2)$$

where N_j is the number of hits observed in time residual bin j and $P(b_j|0\nu)$, $P(b_j|C)$ are the probabilities of observing a hit in time residual bin b_j, for $0\nu\beta\beta$ and cosmogenic events, respectively. These PDFs are just the distributions shown in Figs. 5.2, 5.3 and 5.4.

Figures 5.9, 5.10, and 5.11 show the $\Delta \log \mathcal{L}$ distributions for the three event types, calculated using these PDFs. Each distribution was estimated using 40000 events before cuts. The dashed lines are $0\nu\beta\beta$ events, and their colours indicate which cosmogenic PDF was used to calculate $\Delta \log \mathcal{L}$. In each case, the dashed and dotted lines should be compared as they represent the same discriminant, tuned using the same PDFs. As expected, there is essentially no discrimination power for the β^- group but the separation for the $\beta^+\gamma$ and γ classes is improved to around 1 RMS width and 1.5 RMS widths respectively.

Fig. 5.9 $\Delta \log \mathcal{L}$ distributions for β^- dominated decays. Valid fits only, $r < 3.5\,\text{m}$, $2.43\,\text{MeV} < E < 2.60\,\text{MeV}$

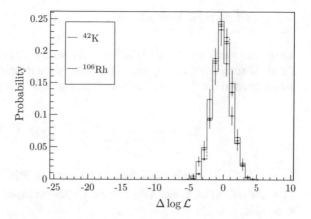

Fig. 5.10 $\Delta \log \mathcal{L}$ distributions for γ dominated decays. Valid fits only, $r < 3.5\,\text{m}$, $2.43\,\text{MeV} < E < 2.60\,\text{MeV}$

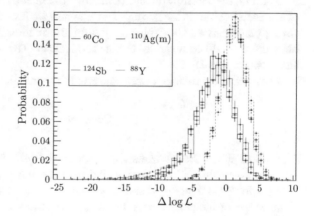

Fig. 5.11 $\Delta \log \mathcal{L}$ distributions for β^+ decays. Valid fits only, $r < 3.5\,\text{m}$, $2.43\,\text{MeV} < E < 2.60\,\text{MeV}$

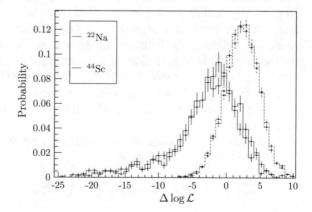

5.1.5 Fisher Discriminant

To go one step further and take into account the correlations between the bins, a Fisher Discriminant can be used. The covariance matrix between all of the bins, Σ, was estimated for $0\nu\beta\beta$, a γ decay, ^{60}Co, and a $\beta^+\gamma$ decay ^{22}Na, according to:

$$\Sigma_{ij} = \frac{1}{N_{ev} - 1} \sum_{\alpha=0}^{N_{ev}} (\tilde{N}_j^\alpha - P(b_j)) \cdot (\tilde{N}_i^\alpha - P(b_i)) \tag{5.3}$$

where N_{ev} is the number of events in the sample, $P(b_j)$ is the binned time residual PDF and $\tilde{N}_j^\alpha$ is content of bin j in the time residual spectrum of event α, normalised to one hit.

Note that the overflow bin has been moved from 200 to 40 ns, to mitigate the effect of low statistics bins in the late tail. The later bins are less well constrained and Fig. 5.1 strongly suggests that later bins carry little information for event classification. The bin index $0 \rightarrow 40$ corresponds to time residuals of $-10 \rightarrow 30$ ns so the prompt peak sits between bins 10–20.

The covariance matrices were converted to correlation matrices according to:

$$\text{Corr}_{ij} = \frac{\Sigma_{ij}}{\Sigma_{ii}} \tag{5.4}$$

Figure 5.12 shows these correlation matrices for $0\nu\beta\beta$, ^{60}Co and ^{22}Na.

There is evidence of structure in all three plots, though the size of the correlations is larger for ^{60}Co than $0\nu\beta\beta$ and larger still for ^{22}Na. The three matrices also differ in the shape of the correlations. For $0\nu\beta\beta$ the only significant structure is negative correlation between bins close to 10, where the prompt peak begins. For ^{60}Co, hits on the rising and falling edge are negatively correlated with hits in the prompt peak and positively correlated with each other. This is the result of variation in γ Compton scattering: a particularly diffuse γ deposit will lead to a late estimation of the event time and extra light on both sides of the prompt peak together. ^{22}Na shows the same structure to a greater extent, as well as additional positive correlations for $j > 20$, resulting from the o-Ps sub-population: if there is lots of light in one late bin, it is likely that o-Ps was formed and therefore there is more likely to be light in the other nearby late bins.

To include these correlations in a discriminant, the Fisher weights $\vec{w}$ for ^{60}Co and ^{22}Na were calculated according to:

$$\vec{w} = (\Sigma_{0\nu} + \Sigma_C)^{-1}(\vec{P}_{0\nu} - \vec{P}_C) \tag{5.5}$$

where $\vec{P}_{0\nu}$ and $\vec{P}_C$ are the bin contents of the normalised time residual spectra. Fisher discriminant parameters for both ^{60}Co and ^{22}Na were then calculated using:

Fig. 5.12 Bin to bin time residual correlations for $0\nu\beta\beta$, ^{60}Co and ^{22}Na. Valid fits only, $r < 3.5\,\mathrm{m}$, $2.43\,\mathrm{MeV} < E < 2.60\,\mathrm{MeV}$

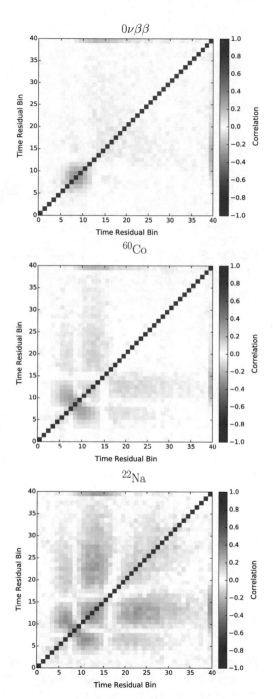

Fig. 5.13 $\mathcal{F}$ for $0\nu\beta\beta$ and ^{22}Na events. Valid fits only, $r < 3.5$ m, 2.43 MeV $< E <$ 2.60 MeV

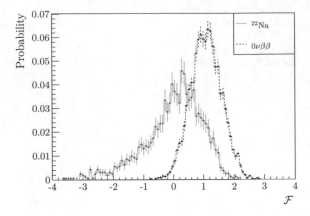

Fig. 5.14 $\mathcal{F}$ for $0\nu\beta\beta$ and ^{60}Co events. Valid fits only, $r < 3.5$ m, 2.43 MeV $< E <$ 2.60 MeV

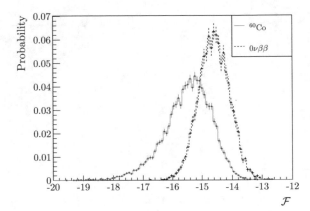

$$\mathcal{F} = \frac{1}{N_{hit}}\vec{N_j}\cdot w \tag{5.6}$$

Figures 5.14 and 5.13 show these two discriminants for ^{60}Co and ^{22}Na against $0\nu\beta\beta$. The ^{60}Co plot shows a slightly greater separation if compared with $\Delta\log\mathcal{L}$ in Fig. 5.10.

5.1.6 Comparison and Power

Figures 5.15 and 5.16 compare the signal and cosmogenic efficiencies that could be achieved by cutting on f_{prompt}, $\Delta\log\mathcal{L}$ and $\mathcal{F}$ for ^{60}Co and ^{22}Na events. In both cases, $\mathcal{F}$ and $\Delta\log\mathcal{L}$ outperform f_{prompt}. The latter two allow for greater background rejections at the same signal sacrifice because they both make use of the full spectral shape of the time residuals, rather than a single bin. For ^{60}Co, including the correlations allows to $\mathcal{F}$ perform better than $\Delta\log\mathcal{L}$ over the whole range. For ^{22}Na

Fig. 5.15 Discriminant comparison for ^{60}Co and $0\nu\beta\beta$ events. Valid fits only, $r < 3.5\,\mathrm{m}$, $2.43\,\mathrm{MeV} < E < 2.60\,\mathrm{MeV}$

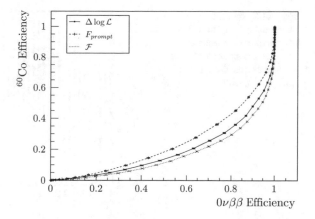

Fig. 5.16 Discriminant comparison for ^{22}Na and $0\nu\beta\beta$ events. Valid fits only, $r < 3.5\,\mathrm{m}$, $2.43\,\mathrm{MeV} < E < 2.60\,\mathrm{MeV}$

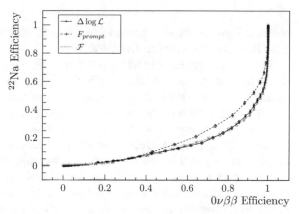

the improvement from using $\mathcal{F}$ is border-line significant, perhaps suggesting that, although correlations are strong, the data are not linearly separable. The two known, distinct sub-populations in ^{22}Na decay (o-Ps and p-Ps) may warrant a more sophisticated approach.

The separation is clearly not large enough for event by event discrimination between e.g. $0\nu\beta\beta$ and ^{60}Co but the discriminants developed in this chapter are useful for *statistically* discriminating between $0\nu\beta\beta$ and background in the case that a large potential signal is observed, a scenario explored in Chap. 8.

5.1.7 Running with E and r

The electron time residuals vary with the energy and radius of the event, so the statistics based on them will too. Figure 5.17 shows the variation of the mean $\Delta \log \mathcal{L}$ as a function of reconstructed energy for $0\nu\beta\beta$ and ^{60}Co events inside the fiducial

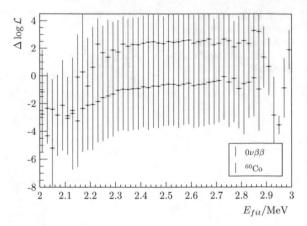

Fig. 5.17 Running of $\Delta \log \mathcal{L}$ for $0\nu\beta\beta$ and ^{60}Co events inside the fiducial volume. Errors are RMS not standard error. Valid fits only, $r < 3.5$ m

volume. Figure 5.18 shows the running with r inside the energy signal window. The error bars show the event to event RMS.[3]

There is a slight upward trend in the statistic with energy for both event types, indicating that the time residuals of both event types look more $0\nu\beta\beta$-like as the energy increases. This is because the reconstructed time residuals become more strongly peaked with increasing energy (shown explicitly in Fig. 3.8) and $0\nu\beta\beta$ is the more peaked hypothesis. However, the separation of the mean $\Delta \log \mathcal{L}$ for the two event types does not change significantly relative to the RMS values, thus the discrimination power is near constant over the range of the $0\nu\beta\beta$ energy window.

Figure 5.18 shows that there is also an upward trend in $\Delta \log \mathcal{L}$ with increasing r, for $r < 5.2$ m. This is because the time residuals become more strongly peaked as the vertex position approaches the AV. For $r > 5.2$ m, there is a sharp downward turn suggesting that both events look more ^{60}Co-like. This is likely to be because of total internal reflection effects of the AV, which make the time residuals less peaked, favouring the broader ^{60}Co hypothesis. Both of these phenomena were shown for the electron time residual PDF in Fig. 3.9.

These two effects show that, if a time residual classifier is used in a likelihood fit for $0\nu\beta\beta$ decay, there will significant variations in its value for both signal and background that must be accounted for to avoid a bias in the fit.

5.2 U/Th Chain

Near the $0\nu\beta\beta$ end point, the significant uranium and thorium chain contributions are ^{210}Tl and ^{214,212}BiPo events. ^{214}Bi, ^{212}Bi and ^{210}Tl are all $\beta\gamma$ decays, so they should have characteristic non-point-like timing signatures that may be used to distinguish them from point-like $0\nu\beta\beta$.

---

[3] Both figures contain 53483 and 19829 ^{60}Co and $0\nu\beta\beta$ events respectively.

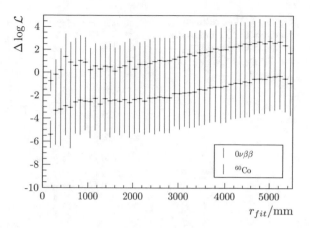

Fig. 5.18 Running of $\Delta \log \mathcal{L}$ for $0\nu\beta\beta$ and ^{60}Co events inside the energy window. Errors are RMS not standard error. Valid fits only, $2.43\,\text{MeV} < E < 2.60\,\text{MeV}$

Unlike the comogenics, these backgrounds are expected to have rates of >1 count/yr but the rates are still a small proportion of the overall background rate, so the goal here is to reduce their count with minimal sacrifice.

5.2.1 Event Topology

BiPo decays where the Bi and Po fall into separate trigger windows are tagged if the two events reconstruct within $\Delta R = 1.5\,\text{m}$ of each other, within a reference time window (see Chap. 4). Of these decays, the ones that are *not* tagged this way are those in which the Bi and Po decays reconstruct a large distance away from each other. J. Wilson showed that this occurs when the Bi decay emits a γ of energy $> 1.7\,\text{MeV}$ [1]. These γ can travel $\mathcal{O}(\text{m})$ before Compton scattering, leading to a reconstructed Bi position that is more than 1.5 m away from the true vertex and the following reconstructed Po event.[4] The PSD technique can be used to reject this residual BiPo background because the high energy γ produced in these events will deposit over multiple sites and create distinctive broad timing profiles.

Events from this part of the decay scheme are very rare. To get the required statistics for this special case, ^{214}Bi events were simulated using a version of RAT 6.1.6 modified to only simulate the parts of the decay scheme that produce a 1.7 MeV γ or greater. 160,000 events were generated for the PDFs and 450,000 events for the test data, before cuts.

^{210}Tl is a β^- emitter with a Q-value of 5.5 MeV. The $\beta\bar{\nu}_e$ pair it produces ranges in energy from 1.4 to 4.4 MeV and there is always the emission of a 0.8 MeV γ. The ^{210}Tl decays which can fall into the $0\nu\beta\beta$ signal window will be those where the

---

[4]Of course, the ΔR cut could be increased to catch these events, but increasing the radius of this cut further leads to unacceptable signal sacrifice: once the near AV region is less than ΔR from the detector centre, the high external event rate leads to many accidental tags.

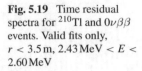

Fig. 5.19 Time residual spectra for ^{210}Tl and $0\nu\beta\beta$ events. Valid fits only, $r < 3.5\,\text{m}$, $2.43\,\text{MeV} < E < 2.60\,\text{MeV}$

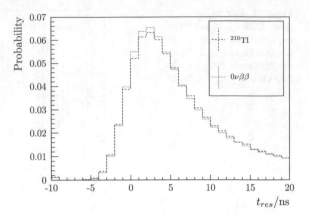

$\bar{\nu}$ carries away at least 2.5 MeV of energy, so these events will have a significant γ energy deposit, and, in principle, they should have broad timing spectra accordingly.

5.2.2   Time Residuals

Figure 5.19 shows an inset of the time residual spectra for $0\nu\beta\beta$ and ^{210}Tl decays. As expected, the ^{210}Tl residuals are broader than those for $0\nu\beta\beta$, but the effect is smaller than for the $\beta\gamma$ and $\beta^{+}\gamma$ decays in Sect. 5.1. This is simply because the 0.8 MeV γ produced in ^{210}Tl decay is lower in energy than many of those produced in the cosmogenic decays.

Figure 5.20 shows the time residuals for ^{214}Bi events which emit γ of 1.7 MeV or more. The events have been divided into several bins of ΔR, defined by the distance between the true and reconstructed vertices. Large ΔR values indicate longer than average γ tracks, which are more likely to be missed by the BiPo coincidence cut. In each case, the ^{214}Bi events are considerably broader than $0\nu\beta\beta$. In particular, there is an excess of early hits at $t_{res} < 0$ which grows with ΔR. These events reconstruct where the γ scatter so the light produced by the β appears to occur before the event.

5.2.3 $\Delta \log \mathcal{L}$

Figure 5.21 shows the $\Delta \log \mathcal{L}$ statistic for ^{210}Tl and $0\nu\beta\beta$ events, calculated relative to the $0\nu\beta\beta$ and ^{210}Tl PDFs in Fig. 5.19. As expected, the separation is slight: the peaks are less than half of one standard deviation apart.

Figure 5.22 shows the same statistic for ^{214}Bi events, for the four bins in ΔR, using the corresponding PDFs in Fig. 5.20. There is a clear trend towards greater separation with increasing ΔR because longer γ tracks produce broader time residuals. Unfor-

Fig. 5.20 ^{214}Bi time residuals for different ΔR values. Valid fits only, $r < 3.5$ m, 2.43 MeV $< E < 2.60$ MeV

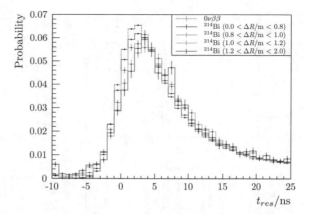

Fig. 5.21 $\Delta \log \mathcal{L}$ for ^{210}Tl events and $0\nu\beta\beta$ events. Valid fits only, $r < 3.5$ m, 2.43 MeV $< E < 2.60$ MeV

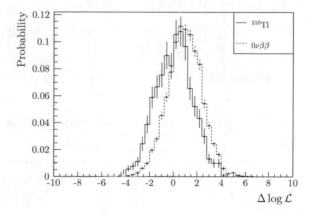

tunately, only a handful of events have $\Delta R > 0.8$ m, so the study is statistically limited.

Figure 5.23 shows the estimated background and signal efficiencies for cuts placed on the $\Delta \log \mathcal{L}$ distributions. The ^{210}Tl cut requires significant signal sacrifice to remove even 10% of events. Given the small contribution of these events to the total background, this cut is not worth making. However, there is power for rejecting ^{214}Bi events, particularly those with large ΔR. For the smallest bin in ΔR, one could reject 60% of background events for around 20% sacrifice. The power of the discriminant improves drastically with increasing ΔR however, because the longest γ paths look most different to point-like $0\nu\beta\beta$ events. Despite the limited statistics, one could conservatively expect to remove 80% of events with $\Delta R > 1.5$ m, without signal sacrifice. High ΔR events are those most likely to be missed by the BiPo coincidence cut, so this is a powerful complimentary technique.

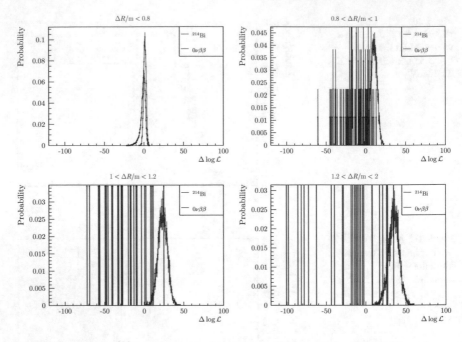

Fig. 5.22 $\Delta \log \mathcal{L}$ for ^{214}Tl events and $0\nu\beta\beta$ events for different ΔR ranges. Valid fits only, $r <$ 3.5 m, 2.43 MeV $< E <$ 2.60 MeV

Fig. 5.23 Efficiency plot for
^{210}Tl and ^{214}Bi backgrounds
and the $0\nu\beta\beta$ signal. Valid
fits only, $r <$ 3.5 m,
2.43 MeV $< E <$ 2.60 MeV

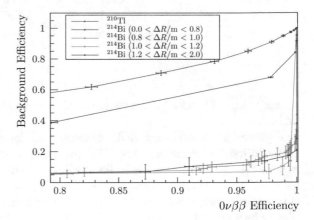

5.3   Calibration

This section explores the robustness of pulse shape parameters for internal back-grounds. It shows that the discrimination power relies only on the best understood part of the time residual spectrum, that a variety of well understood backgrounds may

Fig. 5.24 The effect of moving the underflow/overflow bins on the ^{60}Co analysis. Above: the position of the overflow bins using the full range, and a reduced range. Below: the $\Delta \log \mathcal{L}$ distributions using both ranges. Valid fits only, $r < 3.5$ m, 2.43 MeV $< E <$ 2.60 MeV

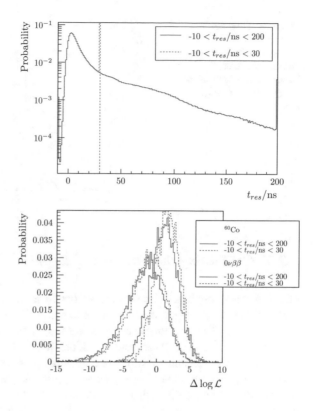

be used to calibrate the technique, and, finally, that only a couple of discriminants are required to reject many different backgrounds.

5.3.1 Overflow Time

Figure 5.1 strongly suggests that the timing differences between $0\nu\beta\beta$ and multi-site backgrounds are concentrated in the prompt peak. This is fortunate as the earliest and latest hits are the hardest to understand: very early hits are dominated by noise and pre-pulsing, whereas very late hits are dominated by absorption-reemission, late pulsing and reflections. In in order that the analysis doesn't depend on the shape of the residuals in these regions, the overflow bins may be moved towards the prompt peak.

Figure 5.24 shows the results of the ^{60}Co analysis where the overflow bin has been moved from 200 to 30 ns, alongside the result using the full 210 ns window. As expected, the results are same up to statistical fluctuations. The Fisher discriminant in Sect. 5.1.5 also used this reduced range.

Fig. 5.25 Time residual
spectra for $0\nu\beta\beta$,
cosmogenic backgrounds of
interest and other
backgrounds which can be
tagged. Valid fits only,
$r < 3.5\,\mathrm{m}$

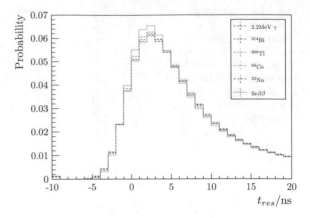

5.3.2   Calibrating the Multi-site Response

Given the small predicted time difference between the point-like and non-point-like
time residual spectra, care must be taken to carefully calibrate the multi-site timing
response inside the detector, so as to reduce the reliance on Monte Carlo.

Fortunately, there are several other well understood backgrounds that may be used
for in-situ measurement. First, a high statistics pure sample of 2.2 MeV γ events
could be extracted by inverting the $(\alpha - n)$ coincidence cut to select the γ created
from neutron capture on hydrogen. Second, with the γ response calibrated, more
complicated $\beta\gamma$ events may be examined by inverting the $^{212/214}$BiPo cuts. There
will be over 4.2×10^5 of these events per year, which can be used to produce a pure,
high statistics sample in the centre of the detector. The broad BiPo energy spectrum
would allow measurement of how the timing response runs with reconstructed energy
and the uniform volume distribution would allow the running with r to be estimated.
Finally, in the fiducial volume, the energy spectrum just above the energy region of
interest is dominated by internal ^{208}Tl decay. This region could be used to gain yet
another sample of $\beta\gamma$ events, this time at a higher energy. Again, the sample can be
purified further using the method of delayed coincidences: there is a characteristic
delay of 3.1 min between the parent ^{212}Bi α decay and the daughter ^{208}Tl $\beta\gamma$ decay.

Figure 5.25 shows the time residual spectra for these 3 event classes, alongside
$0\nu\beta\beta$ and ^{60}Co, ^{22}Na, two cosmogenic isotopes of interest. They are each more similar to one other than to $0\nu\beta\beta$. If the three tagged samples with different energies can
be faithfully reproduced in Monte Carlo, one could reasonably expect to extrapolate
to the particles produced by ^{60}Co, ^{22}Na etc.

To calibrate the positron response one could use the decay of ^{11}C produced by
cosmic muons. These may be tagged using the three fold coincidence technique
demonstrated by Borexino [2]. This would allow for in-situ verification of the o-Ps
lifetime and formation fraction.

Finally, there are ^{44}Sc and ^{60}Co calibration sources [3, 4] that could be used to directly calibrate the $\beta\gamma$ response in the region of interest. These would be best deployed after the observation of a possible signal to avoid contamination risk.

5.3.3 Calibrating the $0\nu\beta\beta$ Response

A similar calibration must be performed for the point-like $0\nu\beta\beta$ response. This can be done with a deployed electron source, or the $2\nu\beta\beta$ background may be used in-situ. Figure 5.26 shows the expected data set for a 3 year live-time in the region 1.6–2 MeV and the fraction of those events that come from $2\nu\beta\beta$. $2\nu\beta\beta$ events contribute over 90% percent of the events collected in this region, with only a small contamination of ^{228}Ac and ^{234}Pa(m) $\beta\gamma$ decays from the thorium and uranium chains, respectively.

The $2\nu\beta\beta$ time residual spectrum differs from $0\nu\beta\beta$ only due to running with energy, so, if the point-like response of $2\nu\beta\beta$ could be extracted in several energy bins, one could use Monte Carlo to extrapolate to the energy of $0\nu\beta\beta$ events. Figure 5.27 shows the gentle variation in the $2\nu\beta\beta$ time residuals with energy and those of $0\nu\beta\beta$.

5.3.4 How Many Discriminants are Needed?

The results of this chapter so far show that, if the timing PDF can be measured, a discriminant can be produced to separate internal backgrounds with γ or β^+ from the $0\nu\beta\beta$ signal, but one discriminant was tuned for each background. Each of these must be calibrated and assigned systematic uncertainties.

The uranium and thorium chain backgrounds are best dealt with using a hard cut with little sacrifice. In these cases, nothing prevents SNO+ using a cut targeted for each background. However, for the cosmogenics, the expected rates are too low and the distributions insufficiently different for a hard cut. Instead, these discriminants are best used as PDFs for $0\nu\beta\beta$ signal extraction. In that case, tuning a discriminant for each of the many internal backgrounds would lead to very high dimensional signal extraction PDFs whose estimates would require prohibitively large Monte Carlo datasets.

Fortunately, the shared event topology of decays in the β^+ and γ groups means that it is sufficient to have one discriminant for each group. Figure 5.28 shows the $\Delta \log \mathcal{L}$ statistic for the γ and $\beta^+\gamma$ cosmogenic isotopes. Each of the γ group isotope likelihoods were calculated using the ^{60}Co PDF, whereas each of the $\beta^+\gamma$ likelihoods were calculated using the ^{22}Na PDF. These are indistinguishable from Figs. 5.3 and 5.4 where each isotope was calculated using its own PDF. This demonstrates that similar topology leads to exact interchangeability from the point of view of PSD.

One might go a step further and ask if it is sufficient to use only one discriminant for both γ and $\beta^+\gamma$ groups, e.g. using the ^{60}Co PDF for both. Figure 5.29 shows

Fig. 5.26 Top: expected
energy spectrum after 1
year's live-time. Bottom:
expected $2\nu\beta\beta$ counts as a
fraction of total counts. Error
bars show expected
variation. BiPo coincidence,
$(\alpha - n)$ coincidence,
$r < 3.5\,\mathrm{m}$ and ITR cuts
applied

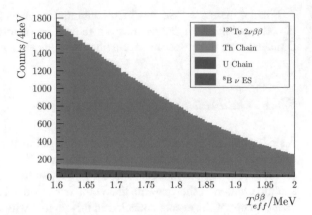

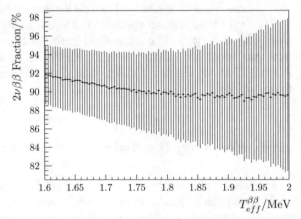

Fig. 5.27 Variation of the
$2\nu\beta\beta/0\nu\beta\beta$ reconstructed
time residuals with energy.
Valid fits only, $r < 3.5\,\mathrm{m}$

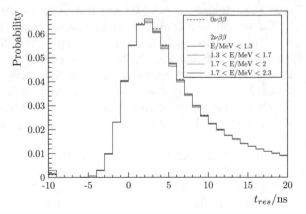

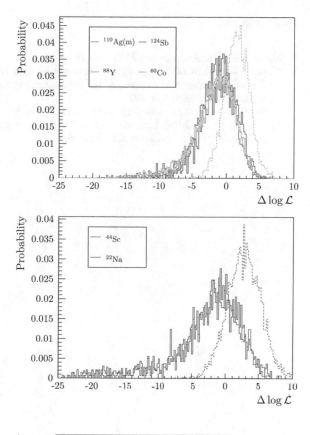

Fig. 5.28 Interchangeability of PDFs within the β^+ and γ event classes. In the first plot the ^{22}Na PDF was used to calculate the likelihood, in the second the ^{60}Co PDF was used. Valid fits only, $r < 3.5$ m, 2.43 MeV $< E < 2.60$ MeV

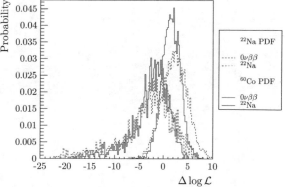

Fig. 5.29 $\Delta \log \mathcal{L}$ for β^+ decays, using ^{60}Co and $0\nu\beta\beta$ time residual PDFs. Valid fits only, $r < 3.5$ m, 2.43 MeV $< E < 2.60$ MeV

the $\Delta \log \mathcal{L}$ statistic for ^{22}Na events using the ^{60}Co PDF (solid line) and ^{22}Na PDF (dashed line). The curves calculated using the ^{60}Co PDF show useful separation, but the performance is better using the correct ^{22}Na PDF, indicating that there is a significant advantage to using a tailored $\beta^+\gamma$ discriminant.

References

1. Wilson J, BiPo rejection update, SNO+-docDB 3881-v3
2. Franco D, Borexino Collaboration (2007) Measurement of the cosmogenic 11C background with the Borexino counting test facility. In: AIP conference proceedings, vol 897(1), pp 111–116. https://doi.org/10.1063/1.2722077
3. Peeters S, Source and interface list, SNO+-docDB 1308-v9
4. Andringa S et al (2016) Current status and future prospects of the SNO+ experiment. arXiv:1508.05759v3 [physics. ins-det] 28 Jan 2016

Chapter 6
Pulse Shape Discrimination of External Backgrounds

The next, more exploratory, chapter investigates the possibility of discriminating between $0\nu\beta\beta$ signal events and radioactive backgrounds created by ^{208}Tl decay outside of the scintillator volume. Of these, ^{208}Tl decay inside the AV, hold-down ropes (HDR), external water (H_2O) and the PMTs (PMT $\beta-\gamma$) form the biggest background contributions inside the fiducial volume, so these are the focus of the work. The analysis updates earlier work by I. Coulter and L. Segui, who first applied a likelihood-ratio technique to the time residuals of these events, in two ways: first, by introducing a new discriminant, based on the angular distribution of hits and, second, by showing that the likelihood-ratio can be improved upon by taking into account correlations between time residual bins, using ^{208}Tl AV as a case study.

The first section of this chapter describes the decays in detail. The second and third sections show that the backgrounds have distinctive hit patterns in time and detector angle that may be used to form discriminant to reject them. The final section investigates correlations between these two discriminants and the power in combining them for ^{208}Tl AV events.

For each data set considered, cuts were applied to select only events that reconstructed as valid, within an energy signal window of $2.438 < E/\mathrm{MeV} < 2.602$[1] and within enlarged fiducial volume of $r < 4.2\,\mathrm{m}$. In each case, the events used to tune the classifiers were completely independent from the events used to test them.

6.1 Decay Details

^{208}Tl β decays to ^{208}Pb with a Q value of 4999 keV and a branching fraction of 100% (Fig. A.1 of Appendix A). The β decay is always to the second excited state of ^{208}Pb, or a higher energy level. All decays reach the first excited state through γ decay, which then decays to the ground state, emitting a 2.61 MeV γ in all cases.

---

[1]This is $\pm$ one energy resolution from $Q_{\beta\beta}$.

© Springer Nature Switzerland AG 2019
J. Dunger, *Event Classification in Liquid Scintillator Using PMT Hit Patterns*,
Springer Theses, https://doi.org/10.1007/978-3-030-31616-7_6

Fig. 6.1 ^{208}Tl decay on the
AV

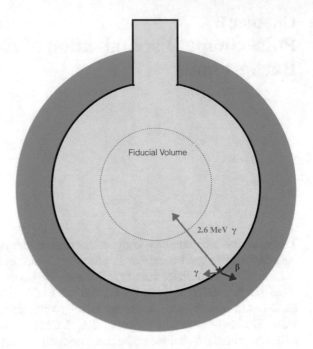

Figure 6.1 shows a cartoon of ^{208}Tl AV decay. The initial β and low energy γ (emitted in decay to the first excited state) deposit their energy close to the AV, but the 2.6 MeV γ can free stream several meters and deposit its energy in the detector centre, mimicking a $0\nu\beta\beta$ event. The same topology is present in all external background events, though the exact location of the initial particles will be different.

Figure 6.2 shows the reconstructed radial position for ^{208}Tl HDR events. Inside the central 4.5 m of the detector, the number of events falls exponentially towards the centre, according to the Compton length of the γ. Only a very small fraction of external events produce a $0\nu\beta\beta$ background, but the rate of decay is large enough to produce a background of a few counts in the signal window per year. The fiducial volume of SNO+ is set at the radius at which the externals cease to be the dominant background. If the external background could be reduced, the optimal fiducial volume would be larger, and more of the detector's volume would be made use of. Therefore, the events of most interest for a rejection study, are those that reconstruct just *outside* the fiducial volume of 3.5 m. This is the reason for choosing a cut of $r < 4.2$ m for the studies that follow.

Two features of this decay can be used to discriminate between these background events and real $0\nu\beta\beta$ events. First, if the β or low energy γ reach the external water, or the scintillator volume, they may create Cherenkov or scintillation light, respectively. This light will produce hits close to the event site that appear very early relative to the main deposition in the centre. Second, the 2.6 MeV γ will deposit its energy over multiple sites, smearing the event's time residuals in the same way as the internal

Fig. 6.2 Exponential
attenuation of external γ
radiation from the hold down
ropes. Points show the
probability that a γ created
by ^{208}Tl decay in the hold
down ropes will reconstruct
at radius r_{fit}, along with an
exponential fit

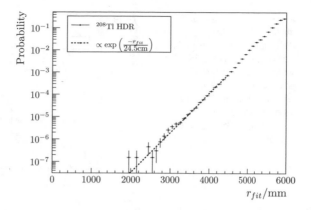

backgrounds in the previous chapter. These two effects give external background
events characteristic signatures in their time residuals and the angular distributions
of their earliest hits. These two handles are explored in the following two sections.

6.2 Timing Discrimination

Figure 6.3 shows the average time residual spectra for $0\nu\beta\beta$ and the four external
decays. The logarithmic plot shows that each of the external backgrounds have an
early hit contribution between -10 and -2 ns that is not present for $0\nu\beta\beta$ events.
The 10–20× larger hit probabilities in this timing window come from the initial γ
and β, which create early hits close to where the decay occurs. The time residuals of
these hits are calculated relative a single reconstructed vertex position at least 4.2 m
away, so these hits look as though they occurred before the event started. The effect
is most noticeable for the AV and HDR ropes, followed by the H$_2$O events, then the
PMT β–γ. This is likely because of the different location of the initial decay: events
on the AV or HDR are close enough that the low energy particles, which deposit
close to the decay, can produce scintillation light in much larger quantities than the
Cherenkov light produced when the particles deposit in the external water. For decays
occurring inside a PMT, it is likely that the early light is emitted at oblique angles
with respect to other nearby PMTs and therefore causes fewer early hits, but further
investigation is required to confirm this hypothesis.

The linear plot shows the second expected difference: for each of the external
backgrounds, the prompt peak is smaller by around 5% and broader on the falling
edge. This is the caused by the multi-site deposition of the 2.6 MeV γ. It creates
several electrons, distributed in space and time, but each time residual is calculated
relative to a single reconstructed vertex, incorrectly accounting for time of flight.
The broadening is close to identical for each of the backgrounds, reflecting the same
2.6 MeV γ produced in each.

Fig. 6.3 Time residuals for $0\nu\beta\beta$ and external background events. Top: the earliest hits on a logarithmic scale. Bottom: the prompt peak on a linear scale. Overflow bins at -10 and 50 ns. Valid fits only, $r < 4.2$ m, $2.438 < E/\text{MeV} < 2.602$

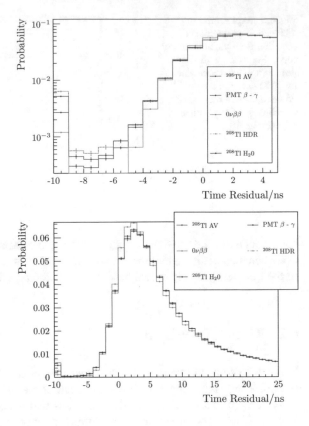

6.2.1 $\Delta \log \mathcal{L}$

Events of unknown origin can be classified as $0\nu\beta\beta$ or external by comparing the event's time residuals against the average distributions in Fig. 6.3, using a likelihood-ratio test. This statistic was calculated in the same way as the internal background discriminant in the previous chapter, using the PDFs shown in Fig. 6.3.

Figure 6.4 shows the $\Delta \log \mathcal{L}$ distributions for the external backgrounds. The $0\nu\beta\beta$ events have a Gaussian distribution in each statistic, consistent with Poisson fluctuations, but, for each of the externals, there are two distinct contributions. Roughly half of the events have a Gaussian distribution, offset from the $0\nu\beta\beta$ peak by around 1/2 of one RMS width. The other contribution is a long tail produced by events that are much more background-like than the former group. This same trend is visible in Fig. 6.5, which shows the number of hits with $t_{res} < -2$ ns for background and signal. For each of the external events, particularly PMT $\beta-\gamma$, there are many events that are consistent with $0\nu\beta\beta$, but, for each, there is an additional tail at high early hit count, which is not consistent with Poisson statistics. Comparisons between the different externals are hampered by poor statistics. Investigations by I. Coulter indicate that these separation contributions could arise from events where the initial β ends its track in the external water, scintillator or acrylic [1].

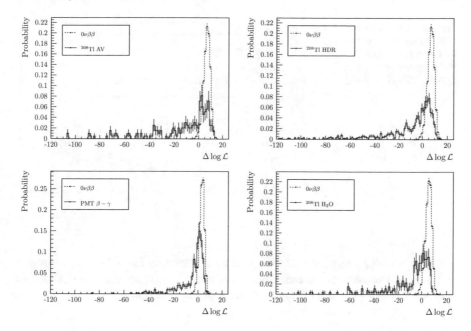

Fig. 6.4 $\Delta \log \mathcal{L}$ distributions comparing $0\nu\beta\beta$ events and external backgrounds. Valid fits only, $r < 4.2\,\mathrm{m}$, $2.438 < E/\mathrm{MeV} < 2.602$

Fig. 6.5 The number of hits with $t_{res} < -2\,\mathrm{ns}$ for $0\nu\beta\beta$ and external background events. Valid fits only, $r < 4.2\,\mathrm{m}$, $2.438 < E/\mathrm{MeV} < 2.602$

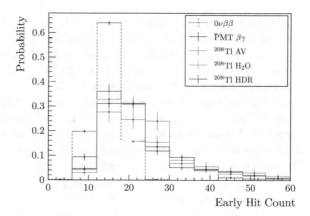

Figure 6.6 shows the background and signal efficiencies that could be obtained by cutting on the $\Delta \log \mathcal{L}$ distributions in Fig. 6.4. At the low signal sacrifice end of the plot, the cut works similarly well for each of the backgrounds, with the exception of the PMT $\beta - \gamma$, for which the performance is noticeably worse. One could expect to reject 50% of AV, HDR and H$_2$O events and 40% of the PMT $\beta - \gamma$ events, with negligible sacrifice.

At larger signal sacrifices, there is larger variation between the different backgrounds but the limited statistics warn against over-interpretation. In particular, the

Fig. 6.6 Efficiencies and
background rejection factors
for the $\Delta \log \mathcal{L}$ statistic.
Valid fits only, $r < 4.2\,\mathrm{m}$,
$2.438 < E/\mathrm{MeV} < 2.602$

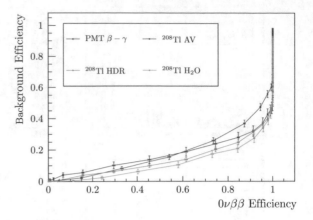

trends in Sect. 6.2.1 suggest that there was a fluctuation in the AV events at around
$\Delta \log \mathcal{L} = 0$ that reduces the estimated rejection and a fluctuation in the H_2O events
at $\Delta \log \mathcal{L} = -5 \rightarrow 0$, which increases the apparent rejection.

6.2.2 Correlations

Evidence for non-Poissonian variation within the external events indicates that there
will be strong correlations between the bins and, therefore that, in some cases, one
could improve upon $\Delta \log \mathcal{L}$, which assumes that the hits are independent. Figure 6.7
shows the bin to bin correlations explicitly for ^{208}Tl AV and $0\nu\beta\beta$ events. The
structure is markedly different between the two.

For the externals, three distinct regions are apparent: the earliest hits in the first
8 bins; the prompt peak in bins 8–15 and the falling edge in bins >15. The earliest
and latest hits are positively correlated with one another, and negatively correlated
with the prompt peak. This suggests that the inherent variability in these events is
between those with lots of light in the prompt peak and those where the prompt peak
is relatively weak, with additional very early and late light.

The $0\nu\beta\beta$ matrix has only one, much weaker, feature. The bins inside the prompt
peak are negatively correlated with one another, and bins on the falling edge are
negatively correlated with later times. This structure is likely caused by variation of
the time residual spectra with event radius (Fig. 3.9).

The interpretation of the external background correlations is hampered by reconstruction effects. The amount of early light and the time delay between it and the
central energy deposit vary between events, but the events are always reconstructed
so that the prompt peak of the time residual spectrum is where it should be for an
electron event. This leads to distortions which are often difficult to explain conclusively. An alternative approach, first employed by L. Segui and K. Majumdar, offsets
the time residuals of each event to the 10th earliest residual in the event:

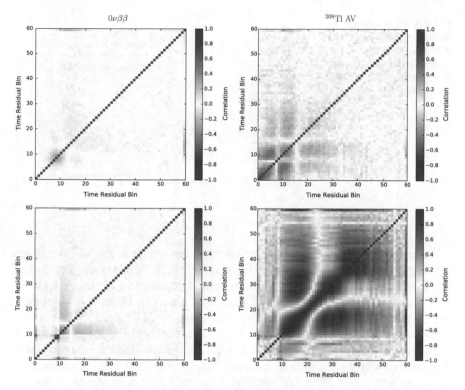

Fig. 6.7 Time residual correlations for $0\nu\beta\beta$ and ^{208}Tl AV events. Above: using t_{res}. Below: using t_{off}. Valid fits only, $r < 4.2$ m, $2.438 < E/\text{MeV} < 2.602$

$$t_{off}^i \rightarrow t_{res}^i - t_{res}^{10} \tag{6.1}$$

where t_{10} is the 10th time residual. This has the effect that all events appear share a common start time, rather than a shared prompt peak. For true electron events, this inevitably introduces jitter into the time residuals, because, for these events, the prompt peak is the best estimate of when the event occurred. But, for external background events, it makes the variable time delay between event start and deposition in the centre apparent.

Figure 6.8 compares t_{res} and t_{off} for $0\nu\beta\beta$ and ^{208}Tl AV events. For $0\nu\beta\beta$, using t_{off} shifts the spectrum to later times (of course) and slightly broadens the spectrum, reflecting jitter in the earliest hits. For ^{208}Tl events, the difference is more drastic: the prompt peak is no longer electron like, because the average spectrum is the sum of events which contain a spectrum of time delays between the initial deposit and the central deposit. This is clear from the t_{off} correlations for ^{208}Tl AV in Fig. 6.7, early and very late hits are strongly anti-correlated with the prompt peak and strongly correlated with each other.

Fig. 6.8 Time residual spectra for $0\nu\beta\beta$ and ^{208}Tl AV background time residuals, raw and offset to the 10th hit. Valid fits only, $r < 4.2\,\mathrm{m}$, $2.438 < E/\mathrm{MeV} < 2.602$

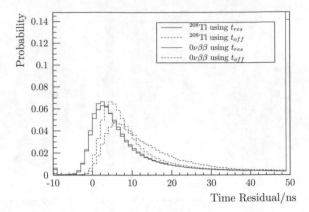

Although it represents the same information, this matrix is simpler to interpret than the correlations for t_{res}. Events with very little early light appear electron-like, because the event appears to start with the central deposit. In these events all of the bins around the electron prompt peak (bin 10) receive more light than average, so these bins are positively correlated with one another.

For events with more early light, the event appears to begin with the initial deposit, and the central deposit arrives some time later. The time delay between the two signals determines where the central peak appears in the spectrum and correlations arise from the variability in that delay. For example, if the main deposit arrives 30 ns after the initial one, bins close to bin 40 will all receive more light than the average for them.

6.2.3   Fisher Discriminant

To investigate the usefulness of these correlations, Fisher discriminants, $\mathcal{F}$, were constructed for the ^{208}Tl AV background. The procedure for constructing the discriminant was exactly as described in Chap. 5, except for the fact that, in this case, the discriminant was constructed using the standard time residuals, t_{res}, and the offset time residuals, t_{off}.

Figure 6.9 shows both discriminants. Unlike the $\Delta \log \mathcal{L}$ statistic, there is no evidence for contributions from distinct groups, rather there is a single smooth curve for the externals. Figure 6.10 compares these two discriminants with $\Delta \log \mathcal{L}$ from the previous section. The first interesting comparison is between the two Fisher discriminants: the offset time residuals appear to perform better over the entire range. This could be because, in t_{off}, the events are more linearly separable, because the correlation matrix is better determined, or simply because there was a statistical fluctuation. The second note is that the Fisher discriminants perform the same to within error, or worse, than the $\Delta \log \mathcal{L}$ statistic close to 100% $0\nu\beta\beta$ efficiency, but significantly out-perform it at the low signal efficiency end.

Fig. 6.9 Fisher discriminants for ^{208}Tl AV and $0\nu\beta\beta$ events, with and without a time residual offset. Valid fits only, $r < 4.2$ m, $2.438 < E/$MeV < 2.602

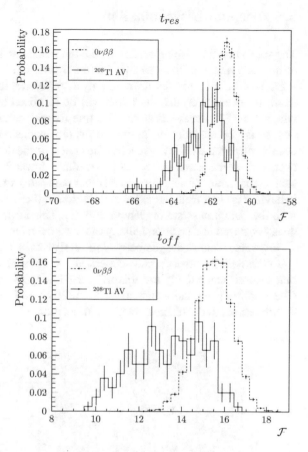

Fig. 6.10 Comparison of timing discriminants for ^{208}Tl AV and $0\nu\beta\beta$ events. Valid fits only, $r < 4.2$ m, $2.438 < E/$MeV < 2.602

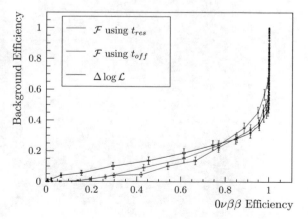

6.3   Angular Discrimination

The second distinguishing feature of the external events is the angular distribution of the early hits. For $0\nu\beta\beta$ events, the hits that do fall into the early window ($t_{res} < -2$ ns) will be noise hits that are randomly distributed across the detector. On the other hand, for external events, those hits will be localised to the region of the detector where the ^{208}Tl decay occurred. The true decay position is, of course, unknown, but it may be statistically inferred from the reconstructed event position. Those events where the 2.6 MeV γ reaches the centre of the detector will be dominated by events where the γ was emitted close to radially inwards (an inward radial path is the shortest path into the detector, and all other paths are exponentially suppressed). This behaviour is clear from the radial distribution of these events, which simply falls off with the Compton scattering length of the γ. This means that, on average, the ^{208}Tl decay occurred at a position radially outward from the reconstructed position.

Figure 6.11 shows the geometry of the early hits in a ^{208}Tl AV event. The 2.6 MeV γ travels radially inwards and scatters in the detector centre to produce an event that reconstructs radially inwards from the initial decay. Associated particles emit Cherenkov or scintillation light to produce early hits. These hits occur at small angles θ with respect to the reconstructed event position, where θ is defined by:

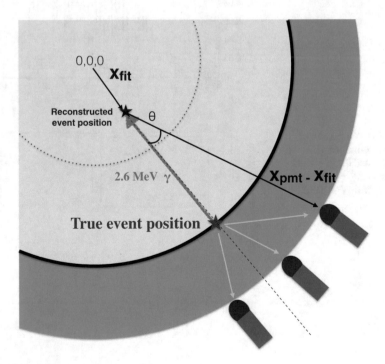

Fig. 6.11   Early hit geometry for external backgrounds

Fig. 6.12 $\cos\theta$ PDFs for $0\nu\beta\beta$ and external background events. Valid fits only, $r < 4.2\,\mathrm{m}$, $2.438 < E/\mathrm{MeV} < 2.602$

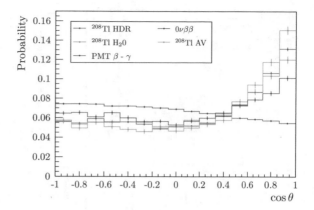

$$\cos\theta = \frac{\mathbf{x}_{fit}}{|\mathbf{x}_{fit}|} \cdot \frac{\mathbf{x}_{PMT} - \mathbf{x}_{fit}}{|\mathbf{x}_{PMT} - \mathbf{x}_{fit}|} \tag{6.2}$$

x_{fit} is the reconstructed event position and x_{PMT} is the position of the hit PMT.

Figure 6.12 shows the $\cos\theta$ probability distributions for hits with $t_{res} < -2\,\mathrm{ns}$, for external and $0\nu\beta\beta$ events. As expected, all four background types show a strong rise towards $\cos\theta = 1$, equivalent to $\theta = 0$. In addition, for both signal and background, there is contribution that rises towards $\cos\theta = -1$, corresponding to PMTs on the far side of the detector. This is the noise hit contribution. Even though noise hits are uniformly distributed across the detector, the time of flight correction makes the furthest away hits appear earliest, so distant noise hits, on the far side of the detector, are most likely to pass the early time cut.

This information can be used to classify events by comparing the $\cos\theta$ distribution observed in an unknown event with the probability distributions in Fig. 6.12 with a likelihood-ratio:

$$\Delta \log \mathcal{L}^{\theta} = \sum_{i=0}^{N_{bins}} N^i \log\left(\frac{P^i_{\beta\beta}}{P^i_{Tl}}\right) \tag{6.3}$$

where N^i is the number of hits observed in $\cos\theta$ bin i, and $P^i_{\beta\beta,Tl}$ are the bin probabilities for signal and background, shown in Fig. 6.12. Figure 6.13 shows this discriminant, comparing each of the four backgrounds with signal. There is evidence of two populations within the external events, as was visible in the timing likelihood-ratio. In each case, the separation is comparable to, but smaller than, than the timing separation achieved in the previous section. Figure 6.14 shows the background and signal efficiencies for cuts on $\Delta \log \mathcal{L}^{\theta}$; 20% of PMT $\beta-\gamma$ events and around 30% of HDR, AV and H_2O events may be rejected with negligible signal sacrifice.

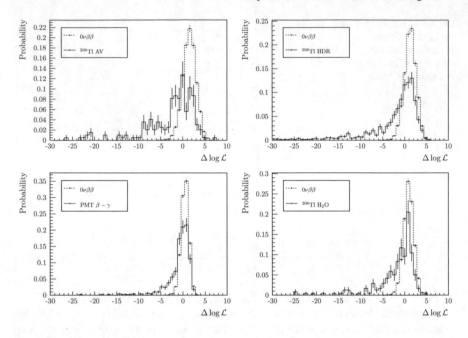

Fig. 6.13 Angular $\Delta \log \mathcal{L}$ for external backgrounds. Valid fits only, $r < 4.2\,\text{m}$, $2.438 < E/\text{MeV} < 2.602$

Fig. 6.14 Rejection efficiency plots for angular $\Delta \log \mathcal{L}$. Valid fits only, $r < 4.2\,\text{m}$, $2.438 < E/\text{MeV} < 2.602$

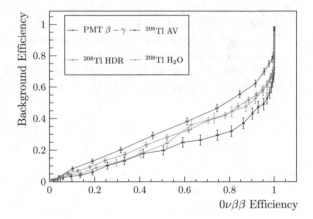

6.4   Combination

The gain from including an angular cut as well as a timing cut depends on to what extent the two discriminants cut the same events, i.e. how correlated they are. Figure 6.15 shows the timing $\mathcal{F}$ discriminant (using t_{off}) against the angular $\Delta \log \mathcal{L}$ for ^{208}Tl (AV) and $0\nu\beta\beta$ events. For background, the two are correlated because both are sensitive to early light from the initial ^{208}Tl: events with little early light

Fig. 6.15 Correlation between angular and timing discriminants for $0\nu\beta\beta$ and ^{208}Tl events. Valid fits only, $r < 4.2$ m, $2.438 < E/\text{MeV} < 2.602$

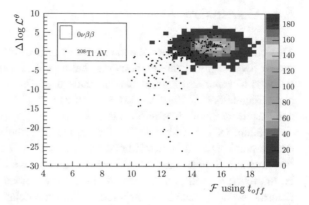

Fig. 6.16 Cut efficiencies for the combination of the timing and angular discriminants. Valid fits only, $r < 4.2$ m, $2.438 < E/\text{MeV} < 2.602$

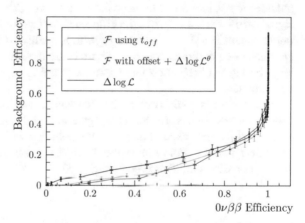

appear electron like in timing, and have random angular distributions in the early light. However, for $0\nu\beta\beta$ events the two are completely uncorrelated: those signal events that happen to have additional early light contributions from noise are no more likely to mimic the angular distribution of an external background. This means that there will be a gain from combining the two discriminants.

The two discriminants were combined using a 2D Fisher discriminant. Figure 6.16 shows the cut efficiencies for signal and background, alongside the timing only discriminant $\mathcal{F}$ and the original $\Delta \log \mathcal{L}$ method for comparison. It shows that adding angular information improves the performance at the high signal efficiency end, but not at high background rejection factors. Using the combined discriminant 90% of background events can be rejected at the cost of 50% signal sacrifice.

6.5   Conclusion

This chapter has confirmed the earlier work of I. Coulter and L. Segui for the most recent SNO+ optical model, showing that a $\Delta \log \mathcal{L}$ statistic, based on timing, can be used to separate external backgrounds from $0\nu\beta\beta$ signal. 40% of PMT $\beta-\gamma$ backgrounds and 50% of the other dominant ^{208}Tl backgrounds within $r < 4.2$ m can be rejected, with negligible signal sacrifice.

In addition, it has shown that the angular distribution of early hits in external events provides further discrimination power, that is partly independent from the first method. Studies on ^{208}Tl AV events also showed that the timing statistic can be improved by taking into account strong correlations between timing bins for the external events, caused by variation in the time delay between the initial deposit outside the AV, and the main deposit in the centre. The correlations buy little at low signal sacrifices, but improve the efficacy significantly at high signal sacrifice. This improvement could be made use of with a likelihood fit which includes the PSD parameter as a fit dimension, or in a hard cut in the outer parts of the detector, where large background rejection factors are required to suppress the dominant external backgrounds.

Further studies could confirm the later conclusions of this chapter with increased Monte Carlo statistics, further investigate the origin of the variability between the events and perhaps expand upon the methods shown here with more sophisticated non-linear methods. For its inclusion in a $0\nu\beta\beta$ likelihood fit, the running with energy and event radius should also be investigated.

Reference

1. Coulter I. Modelling and reconstruction of events in SNO+ related to future searches for lepton and baryon number violation, PhD thesis

Chapter 7
The OXO Signal Extraction Framework

OXO is a C++ signal extraction framework for particle physics with an emphasis on Bayesian statistics. It is written for use in analysis in any particle physics experiment, but its design was informed by SNO+ analyses, particularly the $0\nu\beta\beta$ search. In a nutshell, OXO is a set of C++ classes which represent the major elements of an analysis: the probability distribution functions, parametrisations of systematic uncertainty, test statistics and optimisation or sampling algorithms. Using these components allows for plug-and-play implementations, or they can be mixed with user defined implementations of the underlying interfaces, to achieve a greater level of control.

The code was written mostly by the author of this work but with valuable contributions from others. This chapter briefly reviews it, with particular emphasis on the techniques used for $0\nu\beta\beta$ extraction in Chap. 8.

7.1 Test Statistics

Many particle physics analyses rest on optimising a test statistic with respect to parameters of interest, or sampling the distribution of the statistic with respect to the parameters. Commonly, the test statistic is a likelihood or a chi-squared statistic expressed in terms of fit parameters that represent model uncertainty.

Correspondingly, OXO analyses centre around a `TestStatistic`, an object which returns the value of a statistic for any given set of fit parameters. This can be any C++ class which performs this function; a schematic of the inputs and outputs is given in Fig. 7.1. For SNO+, the statistic might be the likelihood of all the events observed, written in terms of parameters which include the normalisation of a $0\nu\beta\beta$ signal, and the energy resolution of the detector.

Two classes use the input and output of a `TestStatistic` to perform fits. `Optimisers` adjust the fit parameters to find extremal values of the `TestStatistic`, whilst `Samplers` map of the distribution of the statistic with respect to the fit parameters (Fig. 7.2). For optimisation, OXO includes implementations of the grid search algorithm, as well as a wrapper of the ROOT `Minuit`

© Springer Nature Switzerland AG 2019
J. Dunger, *Event Classification in Liquid Scintillator Using PMT Hit Patterns*,
Springer Theses, https://doi.org/10.1007/978-3-030-31616-7_7

Fig. 7.1 Schematic of a `TestStatistic`

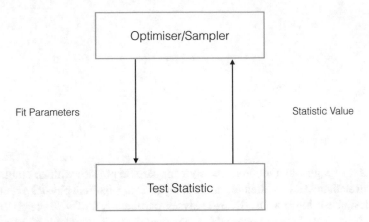

Fig. 7.2 Schematic of an `Optimiser/Sampler`

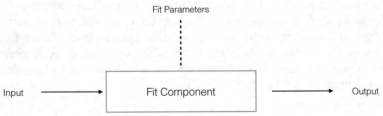

Fig. 7.3 Schematic of a `FitComponent`

Fig. 7.4 The structure of an OXO event

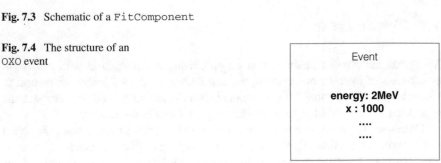

optimiser. For sampling, it offers several Markov chain Monte Carlo algorithms, which are described in detail in Sect. 7.7.2.

A common problem encountered is the need to fit several disjoint data sets simultaneously. For example, one might fit SNO+ data from three separate runs, each with different live-times, tellurium loadings and background rates. To cater for this, any number of OXO test statistics may be combined into a single test statistic using addi-

Fig. 7.5 Schematic of an `EventDistribution`

tion, as would be appropriate for log-likelihoods or chi-squares, or multiplication, for likelihoods. The combination then behaves like a single statistic, on which a single `Sampler` or `Optimiser` can act.

Sometimes parameters must shared between these statistics, for example, the solar mixing parameters should be exactly the same between two SNO+ runs. To achieve this, the parameter need only be given the same name in both statistics for it to be recognised as a single parameter.

The simplest use of the library is to write a `TestStatistic` from scratch, specifying its fit parameters and how to calculate it. Then to apply an `Optimiser` or `Sampler` to it. However, most analyses are expressed in a common language of functions, PDFs, cuts and systematic distortions. Furthermore, many operations on these objects are identical across analyses. For this reason, OXO offers implementations of binned and unbinned PDFs, cuts, systematic distortion of events and distributions etc. that may be built into `TestStatistics`. Often, these operations have associated parameters with uncertainty, which determine the precise action of the operation. For example, an object that performs Gaussian convolutions will have parameters that represent the mean and standard deviation of the kernel. For this, OXO has the concept of a `FitComponent` that allows parameters to be associated with operations (Fig. 7.3). If a `FitComponent` is included in a `TestStatistic`, its fit parameters are automatically recognised by the `TestStatistic` and any `Optimiser`/`Sampler` which acts on it.

The next section describes some of the most fundamental objects, and the features included to solve common problems in particle physics analysis.

7.2 Events and PDFs

Most particle physics experiments read out data in discrete events, each comprising of any number of observables. In OXO, each event is represented as a set of observations keyed by a string containing the name of the corresponding observables (Fig. 7.4).

Another building block of any statistical model is probability distribution functions (PDFs). In particle physics, a particularly important use of PDFs is to represent the relative probability of observing any given event. In OXO, this relationship is expressed by `EventDistributions`, which take in an `Event` and return a probability (Fig. 7.5). Each `EventDistribution` is equipped with a set of observables that determine the dimensionality of the distribution. During any `EventDistribution` operation involving events, the relevant observables for each `EventDistribution` are automatically extracted from each event and, in fits, the normalisation of each PDF included in an OXO fit is automatically recognised as a fit parameter to be optimised.

The most important type of `EventDistribution` is the `BinnedED` class, which represents the internal probability distribution as a binned histogram. This is the most common representation of probability in particle physics, as the distributions are often estimated from Monte Carlo events. However, the use of these distributions is always limited by the curse of dimensionality. Any experiment may produce a large number, n_{dim}, of high level parameters that are powerful for differentiating between different signals. In many cases, producing a full n_{dim} dimensional PDF for signal extraction is infeasible with the available statistics.

OXO offers three approximations for this common problem. First, one can choose to use an analytic PDF if the shape of the distribution is well known. There is complete freedom for users to specify this with any C++ class and uncertainty on the shape of the PDF may be parametrised with any number of fit parameters. This could be used, for example, to describe the ^{8}B ν ES spectrum, floating the solar mixing parameters in the fit. Second, there is functionality for automatically converting any analytic PDF into a binned one, by integrating the analytic distribution over each of the bins. This is particularly useful for maintaining the same code path for the analytic approximation and binned PDFs that may be available with more Monte Carlo in the future. Finally, one can ignore correlations between variables, by replacing a single n_{dim} PDF with the product of several lower dimensional PDFs. For example, for a fit extracting the $0\nu\beta\beta$ rate from SNO+ data, a PSD parameter, C, might only have very weak correlation with reconstructed energy and radius E, r. Then the PDF could be re-factorised as:

$$P(E, r, C) = P(E, r) \times P(C) \tag{7.1}$$

The advantage of this factorisation is that $P(C)$ may be estimated with the same Monte Carlo as $P(E, r)$, so it requires no additional statistics. In OXO, relationships like Eq. 7.1 are expressed simply as a multiplication of `EventDistributions`, which may be binned, analytic or any other form. The result behaves like a single `EventDistribution` in all respects, and, given an event, the correct observables for each are automatically fanned into the respective constituent distributions.

7.3 Propagating Systematic Uncertainty

Often the most robust way of propagating systematic uncertainty is to parametrise the effect of the uncertainty and float the parameters in the fit. This can be achieved in two ways in OXO (both are shown in Fig. 7.6). The first method, `EventSystematic`, models the effect of systematic uncertainty at the event level, by modifying the observed quantities of a single event, with a distortion which depends on the internal fit parameters of the `EventSystematic` object; some examples are given in Table 7.1. These fit parameters are naturally recognised by the `TestStatistic` for inference. Most commonly, these operations are performed on Monte Carlo events

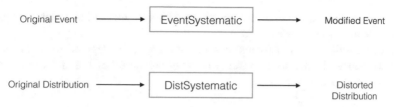

Fig. 7.6 Schematic of an `EventSystematic`

Table 7.1 Example event level systematics. x is the observable in each case, x^T is the true value (only relevant when considering Monte Carlo events). $P(x_1, x_2, x_3..)$ is an arbitrary $N \rightarrow N$ function of any number of observables

Name	Effect	Fit parameters
Scale	$x \rightarrow \alpha x$	α
Shift	$x \rightarrow x + \beta$	β
Gaussian resolution correction	$x \rightarrow x + \gamma(x - x^T)$	γ
Resolution function	$(x_1, x_2, x_3..) \rightarrow$ $P(x_1, x_2, x_3..\|x_1^T, x_2^T, x_3^T..)$	Fit parameters of P

before they are used to fill a template PDF, for example when rebuilding the expected energy spectrum of SNO+ near $Q_{\beta\beta}$, while considering an additional energy shift. The second method models the effect of systematic uncertainty at the probability distribution level, by modifying the bin contents of a `BinnedED` directly. This has the advantage that the effect of any number of these distortions can always be described as a single detector response matrix, M, that relates the undistorted bin contents b_i to the distorted bin contents b_i':

$$b_i' = \sum_{i=0}^{N_{bins}} M_{ij} b_j \tag{7.2}$$

a relation that holds equally well for distributions of any dimension. There are three reasons why, in general, this leads to much faster fits than the first method. First, a single matrix can be calculated to represent *all* systematic distortions, just once, and then applied to any number of distributions for a smaller number of total operations. Second, the number of bins in a typical fit is usually far fewer than the number of events used to build the PDFs. Third, there is highly optimised code available for matrix operations like Eq. 7.2 for both CPUs and GPUs. This method does, however, come with the added risk of bin effects, because information on scales smaller than the bin width is thrown away.

Both methods require care at the fit boundary. Systematics can smear events out of, or in to, the fit region. Simply applying a distortion across the whole range fails to account for the latter effect, leading to a bias. For this reason, OXO has an adjustable buffer region, inside which events are effected by systematics but which are ignored when calculating the test statistic.

A recent addition by B. Liggins allows systematics to be targeted at specific groups of distributions. This is used to treat the energy response of β and α differently in the low energy side-band fit.

7.4   Parameter Constraints

Often there is prior information on the parameters being estimated. Prior measurements of normalisations and systematic uncertainty parameters come from calibration runs, independent side-bands and other experiments. These constraints can pertain to a single parameter, for example a constraint on the ^{8}B ν ES rate from SNO, or relate several parameters, for example a calibration run may provide a correlated constraint on the energy resolution and energy scale. OXO deals with these cases with prior PDFs which can relate any number of parameters. Those PDFs are understood as the prior probability that a given set of values for those parameters is true. It can be user defined or chosen from a number of pre-written alternatives.

7.5   Binned Maximum Likelihood

Several, configurable test statistics come pre-written, including the `BinnedNLLH` class, which implements a standard extended binned log-likelihood of the form:

$$
\log \mathcal{L} = - \sum_{i=0}^{N_{norm}} N_i + \sum_{j=0}^{N_{bins}} N_{obs}^j \log \left(\sum_{i=0}^{N_{norm}} N_i P_i^j(\vec{\Delta}) \right) + \sum_{i=0}^{N_{norm}} C_i(\vec{N}) + \sum_{i=0}^{N_{sys}} \tilde{C}_i(\vec{\Delta})
$$

$$(7.3)$$

where N_i are N_{norm} event class normalisations, N_{obs}^j is the number of events in bin j of the data set, P_i^j is the content of bin j of the `BinnedED` corresponding to normalisation j. The shape of the PDFs are parametrised by a set of systematic variables $\vec{\Delta}$. C_i, $\tilde{C}_i$ are the constraints on the normalisations, $\vec{N}$, and the systematic parameters, $\vec{\Delta}$, respectively. The dependence of the `BinnedED` on the systematic parameters can be expressed using `EventSystematics`, in which case each of the distributions is rebuilt according to $\vec{\Delta}$ at each fit iteration, or `DistSystematics`, in which case the binned distributions are applied directly to the distributions at each step. An adjustable buffer region is included to handle the fit boundary. The corresponding fit parameters are the N_{norm} normalisations and each of the systematic parameters $\vec{\Delta}$. This is the likelihood used for the $0\nu\beta\beta$ fit in the following chapter.

7.6 Datasets

Datasets and histograms can be read from, or written to, ROOT files, or stored in a native HDF5 format.

Most analyses involving fits require a demonstration that the fit parameters are estimated without bias. Doing so requires a number of fake data sets, from which each of the parameters is estimated. The bias and pull of these distributions is a good diagnostic of a well behaved fit. To cater for this, OXO provides utilities for producing fake datasets from binned and unbinned Monte Carlo.

Given unbinned datasets and expected normalisations for each signal type OXO, can split the datasets into a number of fake data sets of user specified live-time, optionally including Poisson fluctuations. The independent remainder can be saved for estimating PDFs or cut efficiencies. Given expected rates and binned or unbinned data, OXO can create the equivalent binned Azimov data set, where each of bin contents is equal to its expected value. This dataset has been shown to obtain the median experimental sensitivity of a search or measurement [1].

Finally, given an Azimov data set (or any other binned data) OXO can add Poisson fluctuations on each bin to produce fake binned data.

7.7 Bayesian Inference

Bayesian analysis can be performed in OXO by defining a test statistic that returns the posterior probability of a set of model parameters and mapping out the posterior using one of the Markov chain Monte Carlo (MCMC) samplers. With this estimate of the posterior distribution, one can calculate credible intervals for limit setting or measurement with error. These concepts are briefly described in the following sections, referencing the $0\nu\beta\beta$ fits performed in Chap. 8.

7.7.1 Posterior Probability

Whereas classical statistics uses probability distributions to describe relative frequency, Bayesian statistics uses them to express degree of belief. In particular, the posterior distribution describes the probability that a given set of model parameters, $\vec{\mu}$, are the correct ones, having seen the observations in an experiment, $\vec{x}$. It is related to the classical likelihood $\mathcal{L}$ by Bayes' theorem:

$$P(\vec{\mu}|\vec{x}) = \mathcal{L}(\vec{\mu}|\vec{x})\frac{P(\vec{\mu})}{P(\vec{x})} \tag{7.4}$$

$P(\vec{x})$ is the probability of observing data set $\vec{x}$, which contributes only a normalisation. $P(\vec{\mu})$ is the prior distribution, which represents the experimenters degree of belief that $\vec{\mu}$ are the correct values *before* looking at the data. The use of a prior and the form it takes are somewhat controversial topics. The analysis in this work follows the principle set out by Biller and Oser [2], setting flat priors on the observable quantities of interest where other constraints are not available.

There are several popular choices for selecting a best fit from the posterior; the study in Chap. 8 chooses to use the most probable point, the set of parameters with maximum a–posterior probability (MAP). This estimate is identical to the maximum likelihood estimator for a flat prior. In this work it was sufficient to estimate the MAP with the highest probability point encountered during sampling.

The posterior distribution has dimensionality equal to the number of fit parameters n_p, including a potentially large number of nuisance parameters. In order to report on just the parameters of interest, the others are typically integrated out in 'marginalisation'. In a search for $0\nu\beta\beta$, only the normalisation of the $0\nu\beta\beta$ signal is important, so all of the other parameters are integrated out to produce the marginalised posterior:

$$P(c_{0\nu}) = \int d\mu^{n_p-1} \, P(\vec{\mu}|x) \tag{7.5}$$

Experiments typically aim to measure, or limit, a quantity to some degree of certainty. A credible interval is a range of parameter values which is believed to contain the true value at a reported degree of confidence. The degree of confidence is simply determined by the integral of the posterior over that interval.

To set a 90% limit on the rate of $0\nu\beta\beta$ events, one simply needs to find the value of $c_{0\nu}$ for which $\int_{-\infty}^{c} P(c_{0\nu}) = 0.9$. Measurements with error are constructed using an interval around the most probable point, which contains 0.683 total probability, equivalent to 1 standard deviation. There are an infinite number of these intervals that may be drawn for any posterior, this work employs the shortest possible interval, constructed by adding points to the interval in descending order of posterior density, until the total probability is reached. Figure 7.7 shows $P(c_{0\nu})$, extracted from two fits in Sect. 8.6. The first has been used to construct a 90% limit, the second to construct a measurement with 1σ error.

7.7.2 Markov chain Monte Carlo

At any given point, it is trivial to calculate a quantity proportional to the posterior $\mathcal{L}(\vec{\mu}|x) \cdot P(\vec{\mu})$, but the normalisation $P(\vec{x})$ cannot be known without mapping out the entire space. When fitting for tens to hundreds of parameters, the curse of dimensionality means that performing this mapping with a simple grid search is intractable.

One could produce samples from P by simply throwing random values of $\vec{\mu}$ and accepting them with probability $P(\vec{\mu})$. However, this becomes extremely inefficient in high dimensional spaces. Markov chain Monte Carlo (MCMC) produces samples

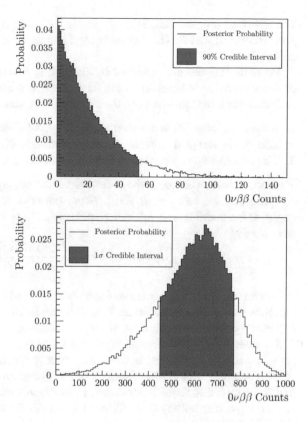

Fig. 7.7 Credible intervals for limit setting and measurement. The posterior estimates are binned, so linear interpolation is performed between bins where necessary

from the posterior in a more efficient way, by algorithmically seeking out regions of high posterior probability.

A Markov chain is a sequence of random events, where the probability of the next event depends only on the current state, not the proceeding events. Provided the chain satisfies certain conditions, it can be shown that the fraction of steps spent in each state will converge on a unique 'stationary' distribution, regardless of the starting point [3].

MCMC treats all of the possible vectors of parameter values, $\vec{\mu}$, as states of a Markov chain. The transition probabilities between them, $T(\vec{\mu} \rightarrow \vec{\mu}')$, are chosen such that the stationary distribution of the chain is the posterior under investigation. In fact, there are infinite choices of $T(\vec{\mu} \rightarrow \vec{\mu}')$. To produce the correct stationary distribution, $T(\vec{\mu} \rightarrow \vec{\mu}')$ must only satisfy the *detailed balance* condition [3], shown in Eq. 7.6.

$$\frac{T(\vec{\mu}' \rightarrow \vec{\mu})}{T(\vec{\mu} \rightarrow \vec{\mu}')} = \frac{P(\mu')}{P(\mu)} \tag{7.6}$$

The transition probabilities, T, are typically factorised into proposal and acceptance probabilities, R and S. The former is used to suggest state transitions, the latter to accept them.

Posterior samples are produced by simulating a Markov chain with the correct stationary distribution and recording its position as it explores the space of possible $\vec{\mu}$. Starting at a random state, $\vec{\mu}_0$, the chain evolves according to:

1. A proposed step, μ', is drawn from the proposal distribution, $R(\mu'|\mu)$.
2. The step is accepted with acceptance probability, $S(\mu'|\mu)$.
3. The current step is recorded as a sample whether a transition occurred or not.

The most common choice is the Metropolis-Hastings algorithm (MH). There, steps are proposed according to $R(\vec{\mu}'|\vec{\mu})$. If the posterior increases over the step, it is accepted unconditionally; otherwise, there is a finite probability of accepting the step anyway:

$$S(\mu'|\mu) = \min\left(1, \frac{P(\mu')}{P(\mu)} \frac{R(\vec{\mu}|\vec{\mu}')}{R(\vec{\mu}'|\vec{\mu})}\right) \tag{7.7}$$

R is often chosen as a Gaussian of fixed width centred on μ. Unfortunately, the MH algorithm becomes very inefficient in high dimensional space with strong correlations between $\vec{\mu}$ components [4]. Such strong correlations naturally arise where there are ambiguities in the data. For example, the following chapter shows that ^{60}Co and $0\nu\beta\beta$ are highly degenerate, because their energy spectra are very similar.

The Hamiltonian MCMC (HMC) method [4] improves on the poor performance of MH on high dimensional posteriors. Rather than a random walk, it proposes new steps using a simulation of Hamiltonian dynamics for a particle in a potential well defined by the posterior.

A classical system with position coordinates q_i and momentum coordinates p_i will evolve according to Hamilton's equations:

$$\frac{dq_i}{dt} = \frac{\partial H}{\partial p_i} \tag{7.8}$$

$$\frac{dp_i}{dt} = -\frac{\partial H}{\partial q_i} \tag{7.9}$$

HMC applies this to the dynamics of a Markov chain by treating the $\vec{\mu}$ states as the position coordinates of a Hamiltonian system. To produce dynamics, fictitious 'momentum' variables are randomly chosen and discarded at each MCMC step. The motivation for this change of view point is that the Hamiltonian, H, is conserved at every point in the trajectory. By treating the posterior as the system's potential energy (Eq. 7.10), a simulation of Hamiltonian dynamics can be used to propose large transitions to points with comparable posterior probability.

$$H = -\log P + \sum_{i=0}^{d} \frac{p_i^2}{2m_i} \tag{7.10}$$

here the 'masses', m_i, are arbitrary parameters that are tuned for efficiency.

The HMC sampling algorithm may be summarised as follows:

1. The position coordinates are set to the current MCMC step parameters, $q_i = \mu_i$.
2. The momentum coordinates are randomly sampled from normal distributions with mean = 0, standard deviation = m_i.
3. N_{steps} of time ϵ are used to simulate Eqs. 7.9 using the leap frog discretisation method [4].
4. The final position and momenta of the trajectory $\vec{q}'$, $\vec{p}'$ are proposed as the next step.
5. The step is accepted with probability min $\left(1, \frac{H(\vec{q}',\vec{p}')}{H(\vec{q},\vec{p})}\right)$.
6. The final $\vec{q}$ is recorded as a posterior sample.

It is well known that MH methods produce slow exploration in the presence of strong correlations between inference parameters. The proposal distribution is typically specified independently for each parameter, so, to achieve a reasonable acceptance rate, it must be very narrow for both of the correlated parameters, leading to slow exploration in the degenerate direction. HMC does not have the same issue: correlations are encoded in the gradient used to create its proposals, so the algorithm automatically accounts for them. For very high dimensional problems, this advantage can easily outweigh the additional calculations required in each HMC step; a detailed discussion with examples is given in [4].

Hard boundaries close to the best fit point can severely damage the efficiency of both HMC and MH. For MH, a proposal distribution that is wide enough to efficiently explore the allowed proportion of the space will frequently propose steps on the other side of the boundary that are rejected. Similarly, the HMC algorithm is designed to propose new points a large distance away from the current point, but if these trajectories frequently cross the boundary, the acceptance probability will be very low. This is significant in a rare process experiment, which may estimate many normalisations, each close to a physical boundary at 0.

The Reflective Refractive HMC (RHMC) method deals with this issue in the HMC algorithm, by modelling reflections of the Hamiltonian trajectories at these boundaries [5]. If a boundary will be crossed by a leap frog step, the step is shortened to take the trajectory to the boundary, the associated momentum is flipped and the remainder of the step is simulated.

OXO includes full implementations of the MH and RHMC samplers for use on any test statistic. The RHMC sampler is used for $0\nu\beta\beta$ signal extraction in the following chapter. Figure 7.8 shows the quality of the samples it produces from a 1D Gaussian with and without a hard boundary at $x = 0$.

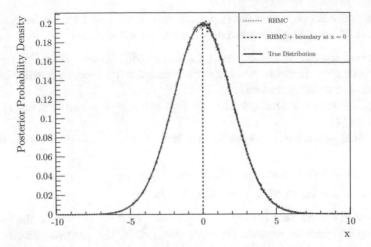

Fig. 7.8 RHMC samples from a $(\mu, \sigma) = (0, 2)$ Gaussian, with and without a boundary at 0

Fig. 7.9 Posterior
auto-correlation versus lag
for the limit setting fit in
Sect. 8.6. There is evidence
of structure over 10s steps,
the fit was run for 50,000
steps

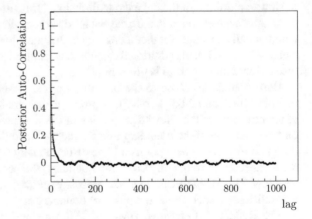

7.7.3 Convergence

The samples drawn from the RHMC algorithm are mathematically guaranteed to
converge to the target posterior in the limit $n_{\text{samples}} \to \infty$. But, for small sample
sizes, the draws are strongly correlated because the chain takes a finite time to explore
the space and reach its stationary distribution. To help asses this convergence, OXO
monitors the auto-correlations between samples, for each of the fit parameters and
the posterior, as a function of the number steps that separate them.

Figure 7.9 shows the estimated auto-correlations of the posterior for the chain
used to produce the $0\nu\beta\beta$ limit in Sect. 8.6. To ensure good mixing, each of the fits
in the following chapter was run for at least 100x longer than the largest features
present in the auto-correlation curve.

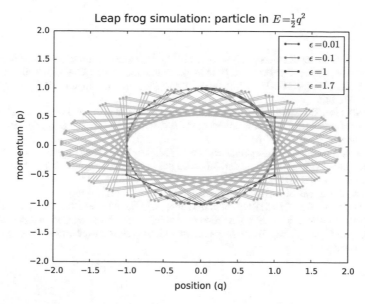

Fig. 7.10 100 steps for a leap frog simulation of a particle with $E = \frac{1}{2}q^2$ for a range of step sizes, ϵ, starting at $q = 1$, $p = 0$. The correct trajectory is a circle of radius 1

The very first steps are particularly correlated, as the random initial position, $\vec{\mu}_0$, will not, in general, be a region of high probability. For this reason, OXO allows an optional 'burn-in' phase during which the first n_{burn} samples are discarded. In this work, 1000 samples was found to be sufficient for the $0\nu\beta\beta$ fit.

The number of iterations required for convergence is acutely sensitive to free parameters in the RHMC proposal distribution. Larger ϵ are more efficient, in general, because the same trajectory may be simulated with fewer steps. However, above a critical value of ϵ, the leap frog discretisation becomes unstable [4]. This leads to incorrect dynamics, biased sampling and very low acceptance probabilities. This behaviour is illustrated in Fig. 7.10. For the smallest two values of ϵ, the dynamics is correctly simulated, and the larger value simulates the trajectory more efficiently. For $\epsilon = 1$ and $\epsilon = 1.7$ the dynamics first becomes oscillatory and then unstable. The $0\nu\beta\beta$ study of this work followed the guidance in [4], selecting the largest value of ϵ that produced sensible acceptance ratios.

With stable dynamics established, N_{steps} determines the length of the trajectory in each iteration. Longer trajectories tend to lead to more independent samples, at the cost of additional computation. Here N_{steps} was chosen as the smallest value that reduced the steady state auto-correlation at a lag of 100 to sub-percent level. The m_i were chosen as roughly the widths of the marginal posterior distributions, estimated in preliminary runs.

References

1. Cowan G, Cranmer K, Gross E, Vitells O (2011) Asymptotic formulae for likelihood-based tests of newphysics. Eur Phys J C71:1554 [Erratum: Eur Phys J C73:2501 (2013)]. https://doi.org/10.1140/epjc/s10052-011-1554-0, https://doi.org/10.1140/epjc/s10052-013-2501-z. arXiv:1007.1727
2. Biller SD, Oser SM (2015) Another look at confidence intervals: proposal for a more relevant and transparent approach. Nucl Instrum Methods A774:103–119. https://doi.org/10.1016/j.nima.2014.11.081. arXiv:1405.5010
3. Gilks WR, Richardson S, Spiegelhalter D (1995) Markov chain Monte Carlo in practice. Chapman & Hall/CRC interdisciplinary statistics. Taylor & Francis, Hoboken. https://books.google.co.uk/books?id=TRXrMWY_i2IC
4. Neal RM (2012) MCMC using Hamiltonian dynamics. arXiv:1206.1901
5. Afshar HM, Domke J (2015) Reflection, refraction, and Hamiltonian Monte Carlo. In: Cortes C et al (eds) Advances in neural information processing systems, vol 28. Curran Associates, Red Hook, pp 3007–3015. http://papers.nips.cc/paper/5801-reflection-refraction-and-hamiltonian-monte-carlo.pdf

Chapter 8
$0\nu\beta\beta$ Extraction

The focus of this chapter is the application of the OXO framework to $0\nu\beta\beta$ extraction in SNO+. Fake data sets with a range of assumed $0\nu\beta\beta$ rates are produced and 2D Bayesian fits in event radius and energy are used to estimate the expected $m_{\beta\beta}$ 90% limit and 3σ discovery level. In addition, the $\Delta \log \mathcal{L}$ statistic, introduced to separate $0\nu\beta\beta$ from ^{60}Co in Chap. 5 is introduced as a third fit dimension to drastically improve the 3σ SNO+ discovery potential.

8.1 Data Set Preparation

The work that follows uses the RAT `6.1.2`, `6.1.6`, and `6.3.2` Monte Carlo production data sets. Between them, these include simulations of every expected SNO+ background.

These data were preprocessed to remove $(\alpha - n)$ and BiPo events that will be tagged by the corresponding coincidence cuts. The $(\alpha - n)$ data sets were cleaned using an implementation of the coincidence cut proposed by Coulter [1]. If any two events were separated by less than 1 ms and both reconstructed above 1 MeV, both events were removed.[1] The cut removes all $(\alpha - n)$ background events within the signal window and fiducial volume. In addition, the $^{214/212}$BiPo coincidence cuts were applied to each of the $^{214/212}$BiPo samples by R. Lane [2]. Events were rejected if they were followed by a second trigger within 500 ns, or another event $500 \rightarrow 3936000$ ns later that reconstructs less than 1.5 m away, with energy >1 MeV. Both of these cuts were not applied to the other backgrounds or $0\nu\beta\beta$, because mixed data sets with more than one event type were not available when the study was conducted.

The data for $0\nu\beta\beta$ and each background was split into two independent data sets of equal size. The first half, the PDF sample, was used to build the signal extraction PDFs; the second, the test sample, was used to create fake data sets.

---

[1] The α decay itself is quenched down to below 1 MeV. However, the prompt signal also contains light produced by neutron elastic scattering on protons (Chap. 4).

© Springer Nature Switzerland AG 2019
J. Dunger, *Event Classification in Liquid Scintillator Using PMT Hit Patterns*,
Springer Theses, https://doi.org/10.1007/978-3-030-31616-7_8

Table 8.1 Assumed signals for Azimov data sets

Data set	$T^{0\nu}_{1/2}/10^{25}$ yr	$0\nu\beta\beta$ Counts/yr before cuts
Azimov 0	∞	0
Azimov 1	17.4	24.87
Azimov 2	7.74	55.96
Azimov 3	4.35	99.48
Azimov 4	2.79	155.44
Azimov 5	1.94	223.83
Azimov 6	1.42	304.65
Azimov 7	1.09	397.91

The test samples for $0\nu\beta\beta$ and each background were then used to create a series of binned Azimov datasets, assuming a live-time of 3 years. For each dataset the backgrounds were scaled to the current best estimates, but different signal normalisations were added to each. Table 8.1 shows these signals, each was calculated according to:

$$0\nu\beta\beta \text{ Counts} = \frac{\log(2)}{T_{1/2}} \times N_{\beta\beta} \times t_{live} \tag{8.1}$$

where t_{live} is the experiment's live-time and $N_{\beta\beta}$ is the number of ^{130}Te nuclei in the detector, calculated using:

$$N_{\beta\beta} = \frac{M_{scint} f_{load}}{m_{Te}} \times f_{ab} \tag{8.2}$$

here f_{ab} is the natural abundance of ^{130}Te, f_{load} is the Te natural loading fraction by mass, M_{scint} the scintillator mass and m_{Te} is the average atomic mass of Te. Using $f_{ab} = 0.3408$, $f_{load} = 0.005$, $M_{scint} = 782$ tonnne and $m_{Te} = 127.60 \times 1.67 \times 10^{-27}$kg, gives $N_{\beta\beta} = 6.253 \times 10^{27}$.

This range of input signals is equivalent to the range $m_{\beta\beta} = 0\text{--}200$ meV, using the IBM-2 prediction of the nuclear matrix element (Sect. 8.8). This range covers the worlds leading limit (100 meV from KamLAND-Zen [3]) and the current best limit on tellurium (140–400 meV from CUORE [4]).

8.2 Cuts

In addition to the coincidence cuts, cuts were applied to several high level parameters produced by RAT, listed in Table 8.2.

The energy window employed is significantly wider than the one used in the counting experiment, in order to incorporate low energy data that are useful for

Table 8.2 Cuts applied in the $0\nu\beta\beta$ fit. The final two are classifiers output by RAT, designed to tag one trigger BiPo events [2, 5]. The ITR classifier was described in Sect. 3.2.1

Name	Variable	Passing values
ROI	Reconstructed energy	1.8–3 MeV
FV	Reconstructed radius	<5.5m
FitValid	Fit validity flag	== 1
ITR	ITR classification	0.33–0.45
BiPo Cumulative	BiPo Cumul classification	<0.0003
BiPo LH	BiPo Likelihood214 classification	<16

Table 8.3 $0\nu\beta\beta$ cut efficiencies estimated with the combined PDF and test samples

Cut	Efficiency/%
FitValid	96.66
FV	75.12
ROI	74.64
Bipo LH	74.55
Bipo cumulative	74.21
ITR	74.03
Total	74.03 ± 0.03

constraining the $2\nu\beta\beta$ background. With this approach, $2\nu\beta\beta$ will be discriminated from $0\nu\beta\beta$ using fine energy bins rather than a hard cut.

Similarly, the fit extends out to 5.5 m, far beyond the canonical fiducial volume of 3.5 m. Of course, the high radius data will contain a large number of external background events which mask any potential signal but the same data is useful for constraining those external backgrounds which do make it into the detector centre and intermediate bins will add to sensitivity. The region 5.5 m–6 m has been excluded because events in this part of the detector are subject to optical effects from the AV which are difficult to understand. In a similar way to the $2\nu\beta\beta$ background, the side-band will be used to constrain the external event rate near the detector centre.

These cuts were applied to both the fake data sets and the PDF samples. Table 8.3 summaries the $0\nu\beta\beta$ efficiency to the cuts, applied sequentially. The total efficiency is $\epsilon_{0\nu} = 74.03 \pm 0.03\%$. The overall signal sacrifice should also include the effect of the coincidence cuts. I. Coulter estimated the $0\nu\beta\beta$ sacrifice from the $(\alpha - n)$ coincidence cut as <0.1%. R. Lane found that the signal sacrifice from the BiPo coincidence cut was negligible [2]. Conservatively assuming a 0.1% sacrifice from these cuts, the total efficiency is also $74.63 \pm 0.03\%$.[2]

---

[2]the error on these cuts are not available.

8.3   Fit Parameters

Fitting all of the possible radioactive backgrounds in SNO+ would require 100 s of fit dimensions. This work took a pragmatic approach, including only the most significant backgrounds as free parameters. The number of counts that each background would contribute to the signal region, after cuts, over a 3 year live-time was estimated using:

$$\text{Counts} = \frac{N_{pass}}{N_{gen}} \cdot r_{exp} \cdot t_{live} \tag{8.3}$$

where N_{pass} is the number of events from entire PDF sample that fall into the signal window defined in Table 8.2, N_{gen} is the number of physics events simulated, r_{exp} is the expected background rate and t_{live} is the assumed live-time of 3 years. Every background that contributed more than 0.1 counts to the signal window had its normalisation floated in the fit. ^{60}Co, ^{88}Y, and ^{22}Na were also included because they are expected to be degenerate with the $0\nu\beta\beta$ signal. In total there are 35 free normalisations in the fit, each is labelled with a (∗) in Appendix A.

8.4   PDFs

The observables selected for this study are the ScintFitter reconstructed energy, $T_{eff}^{\beta\beta}$, and the volume corrected radius, r_{corr}, defined in Eq. 8.4. This parameter has the advantage that constant width bins in r_{corr} contain equal detector volume.

$$r_{corr} = \left(\frac{r}{6\,\text{m}}\right)^3 \tag{8.4}$$

The binning in these two parameters is summarised in Table 8.4. The first r_{corr} bin describes the central 3 m of the detector, which dominates the sensitivity of the experiment. One of these 2D PDFs was produced for each background and $0\nu\beta\beta$ using the PDF datasets.

Six of the PDFs were limited by the available statistics. In particular, the energy spectra inside the two inner most r_{corr} bins were poorly determined for each of the external backgrounds because so few of these events reach the detector centre. This problem was also exhibited in the *internal* ^{208}Tl PDF, where the low energy tail of

Table 8.4 Binning of $0\nu\beta\beta$ signal extraction PDFs

Variable	Minimum	Maximum	Number of bins
$T_{eff}^{\beta\beta}$	1.8	3	48
r_{corr}	0	0.77	6

mis-reconstructed events is significant at high r_{corr}, but very small in the detector centre. Statistical fluctuations in signal extraction PDFs lead to fit biases, because combinations of the fluctuating PDFs can be selected by the fit to match fluctuations in the data. To prevent this, approximate PDFs were produced in each case. For ^{214}Bi from the hold up ropes, the PDF was approximated using the ^{214}Bi hold down rope data, of which there was much more available. These two background sources differ only in their placement in z^3 and their distributions were indistinguishable with the available statistics. In addition, for every external background and internal ^{208}Tl, the energy spectra inside the central two r_{corr} bins were smoothed using one iteration of the ROOT method TH1::Smooth; an example is shown in Fig. 8.1.

8.4.1 PSD PDFs

Fitting a positive $0\nu\beta\beta$ signal requires two things: first, the known background contributions must be sufficiently constrained to demonstrate that a statistically significant excess has been observed at $Q_{\beta\beta}$; second, the excess must be positively identified as $0\nu\beta\beta$. PDFs in $T_{eff}^{\beta\beta}$ and r_{corr} are sufficient to constrain the major expected backgrounds around $Q_{\beta\beta}$ but the decay of some cosmogenic isotopes, particularly ^{60}Co, have identical r_{corr} distributions and very similar $T_{\beta\beta}^{eff}$ distributions to $0\nu\beta\beta$. If an excess of events is observed, the estimated ^{60}Co and $0\nu\beta\beta$ normalisations will be highly degenerate.

Chapter 5 demonstrated that the timing signatures of ^{60}Co and $0\nu\beta\beta$ may be used to differentiate between them. In particular, it showed that a $\Delta \log \mathcal{L}$ statistic, calculated using the time residual spectra of ^{60}Co and $0\nu\beta\beta$ events, was powerful for separating $0\nu\beta\beta$ from all γ dominated cosmogenic decays and that it also had power to separate $0\nu\beta\beta$ events from $\beta^+\gamma$ events (though it did not do so as well as a purpose built statistic). For these reasons, the ^{60}Co-$0\nu\beta\beta$ likelihood-ratio (PSD onwards) was selected as a third fit dimension for the discovery potential study.

Section 5.3 showed that this PSD parameter changes significantly with both energy and reconstructed radius. However, there was insufficient statistics to produce full 3D PDFs in $T_{eff}^{\beta\beta}$, r_{corr} and PSD for the studies. Instead, PDFs of PSD were estimated in each of the r_{corr} bins, $P(\text{PSD}|r_{corr})$, using events between 2.3 and 2.8 MeV. This is the energy range that covers the $0\nu\beta\beta$ and ^{60}Co energy windows and the separation between the two was demonstrated to be constant over this range in Sect. 5.3. The bin definitions are given in Table 8.5. Running within this energy range was ignored and bins outside the ROI were assumed to be flat in PSD. The factorisation can be summarised as:

---

[3] the hold down ropes sit around the top of the AV, the hold down ropes around the bottom.

Fig. 8.1 Top: statistically limited PDF estimate for internal ^{208}Tl. Bottom: the approximate PDF produced to replace it

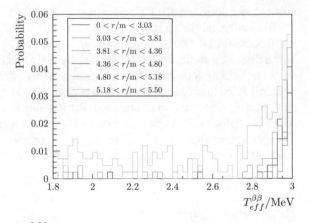

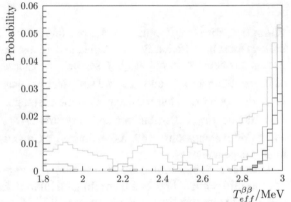

$$P(T_{eff}^{\beta\beta}, r_{corr}, \mathrm{PSD}) = \begin{cases} P(T_{eff}^{\beta\beta}, r_{corr}) \cdot P(\mathrm{PSD}|r_{corr}) & 2.3 < T_{eff}^{\beta\beta}/\mathrm{MeV} < 2.8 \\ P(T_{eff}^{\beta\beta}, r_{corr})/N_{\mathrm{PSD}} & \text{otherwise} \end{cases}$$

(8.5)

here N_{PSD} is the number of PSD bins, included to ensure all events have equal weighting. Factorising the PDFs in this way has the advantage that the studies using them do not rely on knowledge of the $(T_{eff}^{\beta\beta}, r_{corr}, \mathrm{PSD})$ correlations for all 35 signals across the full energy range.

$P(\mathrm{PSD}|r_{corr})$ was estimated directly for $0\nu\beta\beta$, ^{88}Y, ^{22}Na, and ^{60}Co, as well as the four dominant external backgrounds investigated in Chap. 6. For the other backgrounds, appropriate approximations were made. The ^{8}B ν ES background was assumed to look identical to $0\nu\beta\beta$, this is justified by the discussion in Sect. 3.2.2. Sub-dominant external ^{208}Tl backgrounds were assumed to have the same distributions as ^{208}Tl AV. All other backgrounds were conservatively[4] assumed to have the same distribution as $0\nu\beta\beta$. In addition, for the external backgrounds, there was

---

[4]in particular, Sect. 5.2 demonstrated that ^{214}BiPo events in the ROI have a characteristic PSD signature that has not been used here.

Table 8.5 PSD bin definitions

Parameter	Minimum	Maximum	Bin Count
PSD	−14	8	11

insufficient statistics to measure the PSD distributions in the inner most r_{corr} bin. These were assumed to be the same as the the next bin out. This is an unimportant assumption, as the inner most bin is the central 3 m, where the expected external rate is ≪1 count.

Some examples of $P(\text{PSD}|r_{corr})$ are shown in Fig. 8.2. Interestingly, the PSD parameter is able to discriminate between $0\nu\beta\beta$ and external background events, even though it is tuned to spot internal ^{60}Co events.

Exactly the same procedure was performed using the test data to add a PSD dimension to each of the Azimov data sets.

8.5 Posterior

The likelihood in the fit was the standard binned extended maximum likelihood described in Sect. 7.5. The ^{8}B flux at the detector is constrained to $^{+3.62\%}_{-4.00\%}$ by the combined three phase SNO measurement [6]. This information was included in the fit using a Gaussian prior, centred on the expected normalisation (calculated using the procedure described in Sect. 8.3) and a width equal to the average of the positive and negative constraints, $\sigma = 3.8\%$. Note that the uncertainty on the ^{8}B spectral shape and the neutrino mixing parameters have been ignored. Studies by A. Mastbaum suggest these factors contribute a comparable uncertainty to the number of counts in the energy window [7]. All other parameters were given a flat prior, requiring that the normalisation be positive:

$$P_{flat}(c) = \begin{cases} 1 & c \geq 0 \\ 0 & c < 0 \end{cases} \tag{8.6}$$

8.6 Limit Setting

To estimate the expected SNO+ limit on $T^{\beta\beta}_{1/2}$, the fit was performed on the Azimov 0 data set, which contains no $0\nu\beta\beta$ counts. Figure 8.3 shows the fake data alongside the best fit in radial slices. Some of the data, particularly those at high T_{eff}, near the detector centre, show that the Azimov data set suffers from low statistics.

Figure 7.7 showed the $0\nu\beta\beta$ marginal posterior for this fit. At 90% confidence the count limit is 52.6, the error on the total confidence limit from finite sample number

Fig. 8.2 $P(\text{PSD}|r_{corr})$ for
$0\nu\beta\beta$ ^{60}Co and ^{208}Tl HDR.
The first and final bins are
overflows

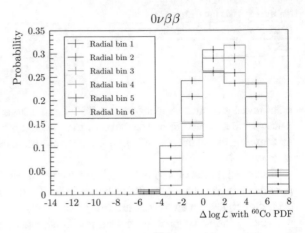

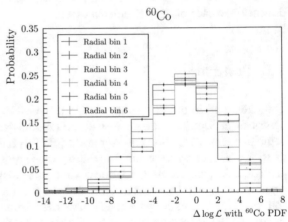

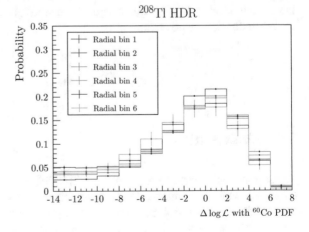

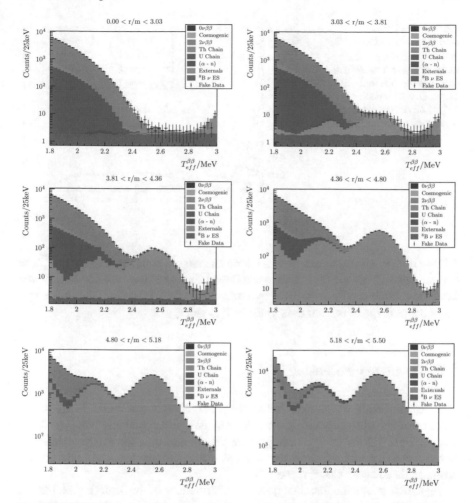

Fig. 8.3 $0\nu\beta\beta$ fit for limit setting in 7 equal volume radial slices. The data points show the 3-year Azimov data set, assuming no $0\nu\beta\beta$ signal; the stacked distributions show the best fit

is 0.1%. The count limit may be converted to a half-life limit according to:

$$T_{1/2}^{90\%} = \frac{\log(2)}{S^{90\%}} \times \epsilon \times N_{\beta\beta} \times t_{live} \qquad (8.7)$$

using signal efficiency $\epsilon = 0.7403$, $t = 3$ yr, and $N_{\beta\beta} = 6.253 \times 10^{27}$ gives $T_{1/2}^{90\%} = 1.76 \times 10^{26}$ yr. This sensitivity is expected to scale proportionally to the square root of the live-time so, after 5 years live-time, one would expect a limit of $T_{1/2}^{90\%} =$

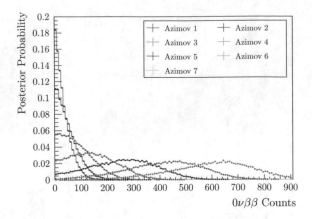

Fig. 8.4 Posterior probabilities extracted using fits in $T_{eff}^{\beta\beta}$ and r_{corr} to a series of 3 year Azimov data sets. The $0\nu\beta\beta$ signal increases from 0 counts to 397.91 counts from Azimov 0 to Azimov 7 (see Table 8.1)

2.27×10^{26} yr, an improvement of 21% on the simple counting experiment.[5] The benefit comes from the fact that the likelihood fit makes use of 4.6× times more detector volume and the entire energy range of the $0\nu\beta\beta$ signal. It also takes into account the shape of each of the signals in $T_{eff}^{\beta\beta}$ and r_{corr}.

8.7 Discovery Potential

This section estimates the expected statistical significance of a $0\nu\beta\beta$ excess as a function of signal size by fitting the 7 Azimov data sets containing non-zero $0\nu\beta\beta$ signals. First 2D fits in $T_{eff}^{\beta\beta}$ and r_{corr} are shown and then the PSD parameter is introduced as a third PDF dimension to break the degeneracy between $0\nu\beta\beta$ and the cosmogenic decays ^{60}Co, ^{88}Y, and ^{22}Na.

Figure 8.4 shows the marginal posterior distributions of the $0\nu\beta\beta$ normalisation, $c_{0\nu}$, for each of the Azimov data sets, only fit in $T_{eff}^{\beta\beta}$ and r_{corr} (without PSD). The most probable $0\nu\beta\beta$ signal increases as the true number of events in the sample increases, as it should. However, the two data sets with the smallest signals are most consistent with 0 signal, and, for each of the other datasets, there is a significant non-Gaussian tail, indicating that even a large $0\nu\beta\beta$ signal is somewhat consistent with $c_{0\nu} = 0$.

Both of these effects are caused by degeneracy between $0\nu\beta\beta$ and the three cosmogenic decays included in the fit. Figure 8.5 shows this explicitly in 2D projections of the posterior for the Azimov 7 dataset that compare $0\nu\beta\beta$ and each of the cosmogenics. The correlation is most significant for ^{60}Co, which shows near maximal correlation, but there is also significant non-Gaussian structure in the ^{22}Na-$0\nu\beta\beta$

---

[5]this improvement is there despite the fact that, without systematic errors, the counting experiment assumes that all of the backgrounds are exactly constrained.

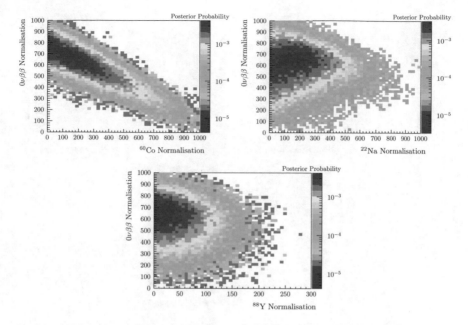

Fig. 8.5 $0\nu\beta\beta$—comogenic posterior correlations without PSD for the fit to the Azimov 7 data set

plane, indicating higher order correlations. This pattern was to be expected, because the energy spectrum of ^{60}Co is most similar to $0\nu\beta\beta$.

To break these correlations, each fit was repeated using datasets and PDFs that also contained a PSD dimension. Figure 8.6 shows the same posterior correlations for the with-PSD fit to data set Azimov 7. They explicitly demonstrate that PSD information breaks the degeneracy between $0\nu\beta\beta$ and the cosmogenics. Figure 8.7 shows the $c_{0\nu}$ posteriors for the with-PSD fits to Azimov data sets 1–7, it shows that the breaking the degeneracy does indeed remove the non-Gaussian tails visible in Fig. 8.4.

8.7.1 Bias

Figure 8.8 shows the most probable $c_{0\nu}$ and 1σ credible intervals for each of the Azimov data sets, fit with and without PSD, and for fits with a 0% constraint on each of the cosmogenics (i.e. the case that their normalisations are known exactly). With a perfect constraint, the fit extracts the correct signal to within error. The small deviations from the 'unbiased' line arise from the finite statistics used to build the Azimov datasets and the PDFs. If the cosmogenics are completely unconstrained, the fit is biased to extract a $0\nu\beta\beta$ normalisation lower than truth by around 200 counts, and signals of less than 200 counts are indistinguishable from zero signal. This is

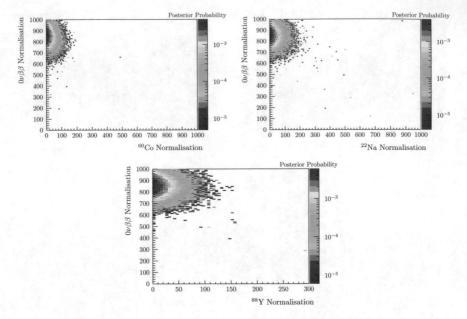

Fig. 8.6 $0\nu\beta\beta$—comogenic posterior correlations with PSD for the fit to the Azimov 7 data set

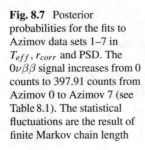

Fig. 8.7 Posterior probabilities for the fits to Azimov data sets 1–7 in T_{eff}, r_{corr} and PSD. The $0\nu\beta\beta$ signal increases from 0 counts to 397.91 counts from Azimov 0 to Azimov 7 (see Table 8.1). The statistical fluctuations are the result of finite Markov chain length

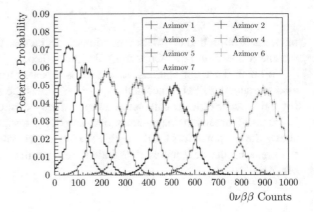

because the fit attributes some of the $0\nu\beta\beta$ bump to cosmogenics. Finally, with PSD, the bias is reduced to around 20 counts, and each point is consistent with the correct $0\nu\beta\beta$ signal to within error.

In summary, degeneracies between cosmogenic decays and the $0\nu\beta\beta$ signal prevent SNO+ from extracting all but the largest signals, unless PSD information is employed, in which case the experiment can measure $0\nu\beta\beta$ signals without systematic bias.

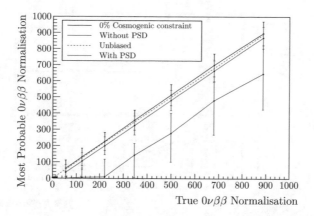

Fig. 8.8 Extracted $0\nu\beta\beta$ signal vs. true signal. The three lines show fits in different observables with different contraints. 'Without PSD' was fit in $T_{eff}^{\beta\beta}$ and r_{corr} alone, floating the normalisations of ^{60}Co, ^{88}Y, and ^{22}Na. '0% Cosmogenic constraint' was produced in the same way, but with the ^{60}Co, ^{88}Y and ^{22}Na normalisations fixed at their true values. Finally, 'With PSD' was produced using fits in $T_{eff}^{\beta\beta}$, r_{corr} and PSD, floating the cosmogenic normalisations. The line labelled 'unbiased' was calculated using the total signal efficiency from Table 8.2

8.7.2 Signal Significance

To claim a discovery of $0\nu\beta\beta$, the no-$0\nu\beta\beta$ null hypothesis must be ruled out to some significance, by asking how inconsistent the $0\nu\beta\beta$ bump is with $c_{0\nu} = 0$. In Bayesian statistics, this question may be answered with the Bayes Factor, but the likelihood-ratio test of classical statistics is equally appropriate and easier to interpret. The likelihood-ratio comparing the two is:

$$\Delta \log \mathcal{L} = \log \left(\frac{\mathcal{L}(c_{0\nu} = 0)}{\mathcal{L}(c_{0\nu} = \hat{c}_{0\nu})} \right) \tag{8.8}$$

where $\mathcal{L}(c_{0\nu} = \hat{c}_{0\nu})$ and $\mathcal{L}(c_{0\nu} = 0)$ are the likelihoods using the best fit and null hypotheses. Wilk's theorem states that, if the null hypothesis is true, $-2\Delta \log \mathcal{L}$ will be χ^2 distributed with 1 degree of freedom in the large sample limit. This distribution can be used to infer the probability of observing a result as, or more, inconsistent with $c_{0\nu} = 0$ as the data set actually observed. The χ_1^2 distribution is equivalent to a Gaussian for $x > 0$ so the statistical significance in standard deviations is simply $n_\sigma = \sqrt{-2\Delta \log \mathcal{L}}$. The fits employed flat priors, so the likelihood-ratio in Eq. 8.8 is exactly the ratio of posterior probabilities between the most probable $c_{0\nu}$ and $c_{0\nu} = 0$. Therefore, for each of the curves in Fig. 8.7 the signal significance is equal to the ratio of the modal posterior probability and the probability of $c_{0\nu} = 0$.

Figure 8.9 shows n_σ for each of the Azimov data sets, against $T_{1/2}^{0\nu}$. Where the statistics were insufficient to evaluate $P(c_{0\nu} = 0)$ directly, a Gaussian fit was used to estimate its value. The PSD fit significantly out-performs the no PSD fit and, for

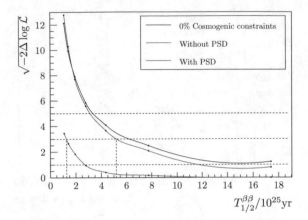

Fig. 8.9 The expected significance with which SNO+ could rule out a $0\nu\beta\beta$ rate of 0, in standard deviation equivalents, as a function of true $0\nu\beta\beta$ half-life. A 3 year live-time is assumed. 'Without PSD' was fit in $T_{eff}^{\beta\beta}$ and r_{corr} alone, floating the normalisations of ^{60}Co, ^{88}Y, and ^{22}Na. '0% Cosmogenic constraint' was produced in the same way, but with the ^{60}Co, ^{88}Y, and ^{22}Na normalisations fixed at their true values. Finally, 'With PSD' was produced using fits in $T_{eff}^{\beta\beta}$, r_{corr} and PSD, floating the cosmogenic normalisations

small signals, performs only slightly worse than if the cosmogenics were perfectly constrained. There is a hint that, for very large signals, the with PSD might actually out-perform the perfectly constrained fit. Though counter intuitive, this feature arises because the PSD statistic also has power to discriminate between $0\nu\beta\beta$ and external background events, which additionally constrains the background normalisation.

Without PSD, one would expect to observe $0\nu\beta\beta$ at 3σ significance if the true half-life is $T_{1/2}^{\beta\beta} = 1.2 \times 10^{25}$ yr. Including PSD, this number is improved by a factor of 4.3 to $T_{1/2}^{\beta\beta} = 5.2 \times 10^{25}$ yr.

8.8 Effective Majorana Mass

$T_{1/2}^{\beta\beta}$ is related to the effective Majorana mass by:

$$\frac{1}{T_{1/2}} = G^{0\nu}||M^{0\nu}||^2 \left(\frac{|m_{\beta\beta}|}{m_e}\right)^2 \qquad (8.9)$$

Table 8.6 shows the nuclear matrix element for $0\nu\beta\beta$ in ^{130}Te, $M^{0\nu}$, as calculated in 5 popular nuclear models. Their combined maximum range is 2.06–4.98. Kotila and Iachello calculated the equivalent phase space factor, $G^{0\nu}$, as 3.688×10^{-14} yr^{-1} with $g_A = 1.269$ [8].

Table 8.6 Range in model predictions for the ^{130}Te dimensionless nuclear matrix element for $g_A = 1.269$ [9–14], taken from [15]

Model	$\|M^{0\nu}\|$
IBM-2	4.03–4.61
QRPA-RU	3.89–4.81
ISM	2.06–2.57
pnQRPA	3.94
EDF	4.98

Table 8.7 Expected SNO+ $m_{\beta\beta}$ limits, assuming a three year live-time. The central values are calculated using the IBM-2 model; the range is calculated using the maximum spread from the predictions of the IBM-2, QRPA-RU, ISM, pnQRPA and EDF models. The first two rows were calculated using fits in $T^{\beta\beta}_{eff}$ and r_{corr} only, the final column was produced using a fit in T_{eff}, r_{corr} and PSD

–	$T^{\beta\beta}_{1/2}/(10^{26}\ \text{yr})$	Central $m_{\beta\beta}$/meV	$m_{\beta\beta}$ Range/meV
90% confidence limit	1.76	49.8	40.3–97.4
3σ Discovery	0.12	191	154–372
3σ Discovery with cosmogenic PSD	0.52	91.5	74.1–179

Table 8.7 shows mass limits, calculated according to Eq. 8.9 for the fits previously described. The ranges have been calculated according to the maximum model range and the central values use the lower limit of the IBM-2 prediction. Including the cosmogenic PSD parameter developed in this work improves the $3\sigma\ m_{\beta\beta}$ sensitivity by 100 meV. Without PSD, the 3σ discovery level is 191 meV, which is already ruled out by the KamLAND-ZEN result [3] and some of the range published by CUORE in ^{130}Te; whereas, with PSD information, this is reduced by 100 meV to 91.5 meV, a signal which is still allowed by all contemporary experiments.

References

1. Coulter I (2017) Alpha-N in Te-Diol, SNO+-docDB 4332-v1
2. Lane R (2017) 0nu counting analysis update with Diol samples, SNO+-docDB 4539-v1
3. Gando A et al (2016) (KamLAND-Zen), Search for Majorana neutrinos near the inverted mass hierarchy region with KamLAND-Zen. Phys Rev Lett 117(8):082503. https://doi.org/10.1103/PhysRevLett.117.109903, https://doi.org/10.1103/PhysRevLett.117.082503, [Addendum: Phys. Rev. Lett.117, no.10,109903(2016)], arXiv:1605.02889
4. Alduino C et al (2017) (CUORE), First results from CUORE: a search for lepton number violation via $0\nu\beta\beta$ Decay of ^{130}Te. arXiv:1710.07988
5. Majumdar K On the measurement of optical scattering and studies of background rejection in the SNO+ Detector. PhD thesis

6. Aharmim B et al (2013) (SNO), Combined analysis of all three phases of solar neutrino data from the sudbury neutrino observatory. Phys Rev C88:025501. https://doi.org/10.1103/PhysRevC.88.025501, arXiv:1109.0763

7. Mastbaum A (2015) Systematics and constraints in the double beta counting analysis, SNO+-docDB 3000-v1

8. Kotila J, Iachello F (2012) Phase-space factors for double-β decay. Phys. Rev. C 85:034316. https://doi.org/10.1103/PhysRevC.85.034316

9. Fedor S, Vadim R, Amand F, Petr V (2013) 0$\nu\beta\beta$ and 2$\nu\beta\beta$ nuclear matrix elements, quasiparticle random-phase approximation, and isospin symmetry restoration. Phys Rev C 87:045501. https://doi.org/10.1103/PhysRevC.87.045501

10. Barea J, Iachello F (2009) Neutrinoless double-β decay in the microscopic interacting boson model. Phys Rev C 79:044301. https://doi.org/10.1103/PhysRevC.79.044301

11. Barea J, Kotila J, Iachello F (2013) Nuclear matrix elements for double-β decay. Phys Rev C 87:014315. https://doi.org/10.1103/PhysRevC.87.014315

12. Menendez J, Poves A, Caurier E, Nowacki F (2009) Disassembling the nuclear matrix elements of the neutrinoless beta beta decay. Nucl Phys A 818:139–151. https://doi.org/10.1016/j.nuclphysa.2008.12.005, arXiv:0801.3760

13. Juhani H, Jouni S (2015) Nuclear matrix elements for 0$\nu\beta\beta$ decays with light or heavy Majorana-neutrino exchange. Phys Rev C 91:024613. https://doi.org/10.1103/PhysRevC.91.024613

14. Rodrguez TR, Martnez-Pinedo G (2013) Neutrinoless decay nuclear matrix elements in an isotopic chain. Phys Lett B 719(1):174–178. https://doi.org/10.1016/j.physletb.2012.12.063, http://www.sciencedirect.com/science/article/pii/S0370269312013160

15. Mastbaum A (2017) Double-beta white paper sensitivity plots, SNO+-docDB 2593-v10

Chapter 9
Neutrinoless Double Beta Decay with Slow Scintillator

A strength of water Cherenkov detectors is that the direction, quantity and isotropy of Cherenkov radiation produced in physics events encodes information about the type and direction of the particle(s) that produced it. In particular, solar neutrino events which produce electrons may be identified using the angle the electron subtends with the solar direction. If a $0\nu\beta\beta$ search was performed in a water Cherenkov detector, this technique could be used to reduce the ^{8}B ν ES background that limits the sensitivity of SNO+. Moreover, if the directionality and energy split of the two electrons emitted in $0\nu\beta\beta$ could be estimated from their Cherenkov signals, this information could be used to determine the underlying $0\nu\beta\beta$ mechanism. However, the relatively modest Cherenkov yield prohibits a ^{130}Te $0\nu\beta\beta$ search in water: in modern water Cherenkov detectors, the $\mathcal{O}(1\,\mathrm{MeV})$ events of interest are poorly reconstructed or below detector threshold altogether.

On the other hand, liquid scintillator detectors have far greater light yields, that allow for the low thresholds and good energy resolution demanded by a $0\nu\beta\beta$ search. However, the emitted light is isotropic and distributed in time, so particle type can only be inferred from the scintillation pulse shape. This is the technique explored in the early chapters of this work. Critically, it has no discrimination power for the limiting ^{8}B ν ES background, nor for distinguishing between $0\nu\beta\beta$ mechanisms.

Of course, in a liquid scintillator detector, both emission mechanisms are at work and, in principle, the Cherenkov and scintillation signals are separable in time. Cherenkov emission is near instantaneous, whereas scintillation photons are emitted only after the decay of excited molecular states with lifetimes of 10–100 s of ns.

The problem is that contemporary scintillation detectors are unable to resolve this time difference. Current liquid scintillation detectors have rise and fall times of <1 ns and 3–5 s ns respectively, so the two signals are typically separated by just a couple of ns. This time difference is washed out by dispersion, timing uncertainties from finite vertex resolution, and the TTS of large area PMTs. The result is that the Cherenkov signal is buried under the much larger scintillation pulse.

Water based liquid scintillators (WbLS) propose to enhance the Cherenkov signal by diluting liquid scintillator with UPW (e.g. [1]). This reduces the overall scintil-

© Springer Nature Switzerland AG 2019
J. Dunger, *Event Classification in Liquid Scintillator Using PMT Hit Patterns*, Springer Theses, https://doi.org/10.1007/978-3-030-31616-7_9

lation yield so that the Cherenkov signal may be more easily identified. For a $0\nu\beta\beta$ experiment, this comes with two major drawbacks. First, by reducing the light yield, one necessarily admits more $2\nu\beta\beta$ background into the $0\nu\beta\beta$ signal window. Second, UPW has $100\times$ the uranium and thorium chain contamination of purified liquid scintillator, so a WbLS $0\nu\beta\beta$ experiment will have to deal with orders of magnitude more internal radioactivity than a pure scintillator detector.

An alternative solution, explored here, is to enhance the timing separation between the two signals instead, using a slow scintillator and a high coverage of fast, high quantum efficiency PMTs. Increasing the scintillator's rise time increases the time difference between the two signals, and increasing its fall time[1] reduces the size of the scintillation peak relative to the Cherenkov pulse. State of the art PMTs with sub-ns timing resolution could better resolve the Cherenkov-scintillation time difference and a larger coverage of high quantum efficiency PMTs would increase collection of the scarce Cherenkov signal. If achieved, one could have liquid scintillator light yields and purifications particle ID using Cherenkov light and in the same detector.

So far, the stumbling block with this approach has been discovering a scintillator with slow timing and a sufficient light output. However, recently several such organic scintillators have been found by S. Biller, with rise times of 1.5–15 ns and fall times of 10–50 ns [2].

This final chapter discusses the capabilities of a SNO+ like detector, filled with one of these scintillators and equipped with a dense coverage of high quantum efficiency, fast PMTs. The scintillator with the longest rise and fall time was selected as a proof of principle. Its exact formula will be named in a separate publication, so here it is referred to as simply 'slow-scintillator'. The reader should be aware that the techniques under current investigation for loading these scintillators with tellurium strongly quench the scintillator light output. Therefore, the work that follows depends on a future solution to this problem.

After describing a RAT simulation of the proposed detector, it is shown that the detector can separate the scintillation and Cherenkov signals in time, and that $0\nu\beta\beta$ and ^8B ν ES events may be fully reconstructed. Differences in the isotropy, directionality and quantity of Cherenkov light produced in these event types are exploited to distinguish them statistically and the possibility of determining the $0\nu\beta\beta$ mechanism is investigated.

9.1 Detector Model

Figure 9.1 shows a schematic of the detector considered. At its centre sits a perfectly spherical acrylic vessel (AV) of 8.8 m radius, 5 cm thickness and filled with liquid scintillator. 11840 mm from the centre of the AV are the front faces of 21873 identical PMTs, pointing radially inwards. UPW fills the space from the AV to the PMTs and the space behind them.

---

[1]This is determined by the life-time of the scintillating excited molecular states.

Fig. 9.1 Schematic of the
detector geometry. AV
⌀ 8.8 m, front faces of the
21873 PMTs lie on a 23.68 m
⌀ sphere. Not to scale

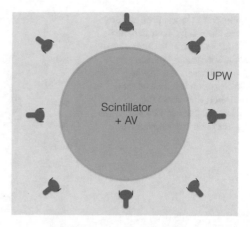

To more realistically model a future experiment, the SNO+ PMTs have been
replaced with the current state of the art technology and the reflectivity of the concentrators has been adjusted so they behave as new.

The detector was simulated using a modified version of SNO+ RAT. This required
a new GEANT4 geometry, a full scintillator model informed by measurements of
slow-scintillator in Oxford and modifications to the behaviour of the concentrators,
PMTs and DAQ system. Each assumption is stated and justified in the following
sections.

9.1.1 Scintillator

The AV is filled with LAB, doped with 10 g/L of slow-scintillator. LABSS will be used
to refer to these two components together. At this concentration, there is non-radiative
transfer of excitations between the LAB and the slow-scintillator. This means that
primary scintillation light is emitted from slow-scintillator and that photons absorbed
by either LAB or slow-scintillator are re-emitted from the slow-scintillator.

The primary advantage of the slow-scintillator is, of course, its timing; it has a rise
time of 16 ns and a single fall time of 50 ns. A comparison of the scintillator emission
times of LABSS and LABPPO is shown in Fig. 9.4. The prompt peak is $>10\times$
smaller with slow-scintillator, so it produces much less scintillation background to
the Cherenkov signal emitted in the first 60 ps (Fig. 2.8).

Figure 9.2 shows the emission and absorption curves for LAB and slow-scintillator,
alongside the PMT efficiency curve employed in the simulation. The slow-scintillator's
emission peak is at 490 nm, the emission distribution matches the PMT efficiency
reasonably well and the cocktail displays a large Stokes' shift.

The absorption lengths for the detector's major optical components are shown in
Fig. 9.3. Slow-scintillator strongly absorbs in the 290–390 nm region. The PMTs are
sensitive in this region so there is significant absorption of Cherenkov photons that

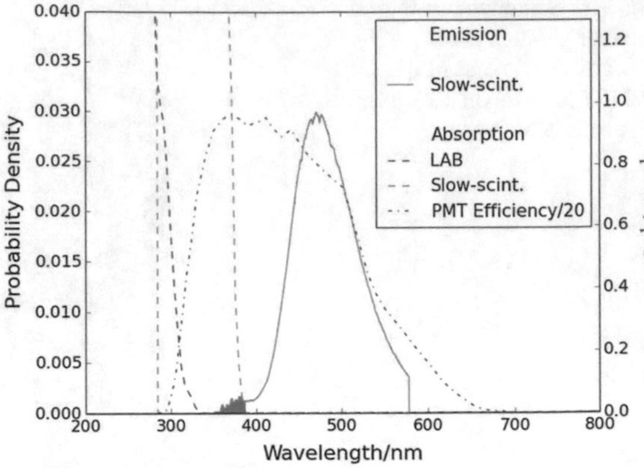

Fig. 9.2 Optical profiles for the slow-scintillator detector components. Solid lines show emission spectra, dotted lines show inverse absorption lengths and the dashed line shows the Hamamatsu R5912 PMT combined efficiency curve

Fig. 9.3 Absorption lengths for slow-scintillator detector components

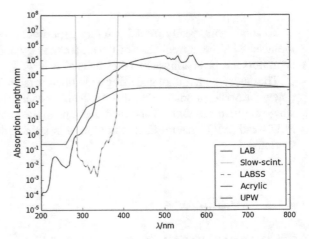

would be otherwise detected. These photons are re-emitted from slow-scintillator, mixing them in with the scintillation signal. The scattering of LABSS was assumed to be the same as LABPPO, because both will be dominated by LAB.

The intrinsic light yield of LABSS is different from LABPPO for two reasons. First, non-radiative transfer from LAB to slow-scintillator is more efficient than the LAB to PPO case (0.835 vs. 0.75). Second, the quantum yield is lower for slow-scintillator than PPO (0.75 vs. 0.8). Overall, the intrinsic light yield of LABSS is 10765 γ/MeV, 10% lower than LABPPO but 60% higher than the quenched 0.5% TeDiol cocktail to be deployed in SNO+ phase I.

The optical effects of the loaded $0\nu\beta\beta$ isotope, assumed to be tellurium, have been neglected. Absorption from tellurium itself is very small even at percent level loadings [3], but the chemistry required to *load* the tellurium is highly uncertain.

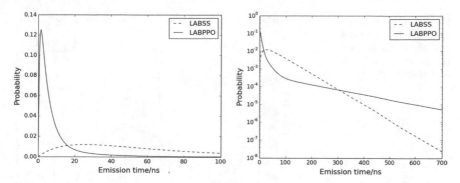

Fig. 9.4 Scintillation photon emission time distributions for LABPPO [4] and LABSS on linear and logarithmic scales

9.1.2 PMTs

The detector is equipped with fast high quantum efficiency PMTs. The exact specifications were chosen to represent roughly the best 8" PMTs on the market. The charge response and dark current of the fast PMTs was assumed to be the same as the SNO+ PMTs (Fig. 2.5). This is a conservative because modern PMTs have a far better ratio between charge pedestal and single p.e. response (e.g. [5]). Using the same front end electronics, as much as 10% more p.e. could be without admitting additional noise hits [6].

Modern PMTs have much better detection efficiencies than the SNO PMTs. In particular, HQE (high quantum efficiency) PMTs are equipped with super bialkali photo-cathodes that increase their quantum efficiency by around 35% relative to those without (e.g. [7]). The PMTs considered here were assumed to have the same efficiency as the Hamamatsu R5912 8' HQE. A trace of its combined efficiency as a function of wavelength is shown in Fig. 9.5, alongside the SNO+ R1408 PMT. In the studies that follow, the R5912 is used to detect light emitted by slow-scintillator with $\lambda \approx 470$ mm, whereas, in SNO+, the R1408 will be used to detect light emitted by bisMSB, $\lambda \approx 420$ nm. Integrated over the relevant emission curves in Figs. 9.2 and 2.10, the R5912 is 62% more efficient than the R1408.

The PMT TTS was assumed to be the same as the R1408, but with a prompt peak of 1 ns FWHM. Under this assumption the pre-pulsing and after-pulsing behaviour is the same as the R1408. The assumed transit time distribution is shown in Fig. 9.6, it was created by modifying the R1408 distribution. A Gaussian was fit to the R1408 prompt peak, this Gaussian was then subtracted and replaced by a 1ns FWHM Gaussian of the same mean and normalisation. Large area PMTs with close to this timing performance are already available. For example, the R5912-MOD has a measured TTS of 1.5 ns [5]. This PMT could also be made into a HQE PMT if equipped with a super bialkali [5].

Fig. 9.5 Comparing the combined (quantum and collection) efficiencies of the Hamamatsu R5912 and R1408 PMTs [5, 8]

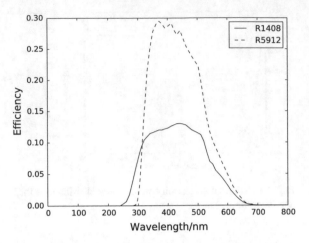

Fig. 9.6 PMT transit time distribution used for the slow-scintillator studies, compared with the transit time distribution of the Hamamatsu R1408

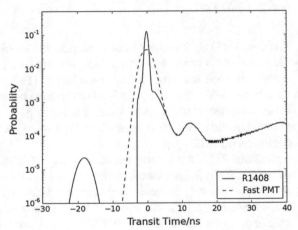

9.1.3 Photo-Cathode Coverage

The geometry of the PMTs was assumed to be identical to the SNO+ PMTs and each one was simulated with a SNO+ concentrator. However, relative to SNO+, the reflectivity of the concentrators was increased from 83% to 91%. This improvement could be achieved by simply replacing the SNO+ concentrators with new ones of the same design. More reflective materials and tessellating concentrators could increase the light collection even further [9].

The centre of each of the 21,873 PMT front faces sits on a sphere of radius 11,840 mm. Their positions were chosen to maximise the number on the sphere without the PMT concentrators overlapping. To achieve this, first a sphere of radius 11840 mm was pixelised using icosahedron subdivision (see Fig. 9.7) and a PMT was placed at the centre of each pixel. Second, any two PMTs lying closer than twice the concentrator radius (28 cm) were removed and mutual repulsion between

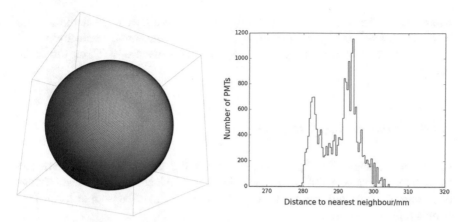

Fig. 9.7 Left: approximation to the sphere using icosahedron division. Right: nearest neighbour distances for the final PMT geometry

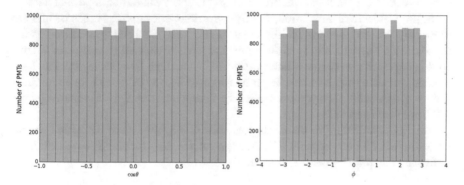

Fig. 9.8 $\cos\theta$ and ϕ of the PMTs relative to the z-axis

all of the PMT centres was simulated to close the 'hole' created and preserve the symmetry.[2] This process was repeated until no two PMTs were separated by less than 28 cm. Figure 9.7 shows the final distribution of nearest neighbour separations; the symmetry of the arrangement is clear from the $\cos\theta$ and ϕ distributions in Fig. 9.8.

In this configuration, the physical coverage is 76.5%. To account for the imperfect reflectivity of the concentrators, it is more useful to estimate the effective coverage:

$$C_{eff} = C_{phys} \cdot ((1 - f) + f R) \tag{9.1}$$

---

[2]The sphere pixelisation was the work of the author, the mutual repulsion code was developed by E. Leming.

where f is the fraction of the concentrator opening area covered by reflectors, $1 - f$ is the fraction covered by photo-cathode and R is the concentrator reflectivity.

$f = 0.5$, $R = 0.91$ gives $C_{eff} = 73\%$, an improvement of 37% on the SNO+ detector [10].

9.1.4 Front End and Trigger

In order to collect all of the light generated by LABSS, the trigger gate was extended from the SNO+ value of 400–600 ns, and the trigger lock out to 610 ns. Each channel discriminator threshold was set to a typical value for the SNO+ detector, 9 DAC counts [10].

The trigger thresholds were left unchanged from the SNO+ values. A real detector would likely raise these thresholds to compensate for greater light collection, but $0\nu\beta\beta$ events will be well above threshold, so the exact value is unimportant here.

In the SNO+ RAT simulation, the true discriminator crossing times of hits are artificially smeared, and then calibrated using PCA/ECA constants measured on the real detector. These calibrations do not extend past the channel reset time of 410 ns, so it is impossible to simulate later hits with this method. In order to simulate later hits, the truth discriminator crossing time was used instead of the calibrated hit time. This is a reasonable approximation because timing uncertainty is dominated by the TTS, even for fast PMTs.

More modern DAQ systems employ waveform digitizers to record the PMT charge pulses in full. If equipped with a similar system, the detector could also distinguish individual p.e. piling up on a single PMT (e.g. [11]).

9.1.5 Detector Performance

In this section we briefly highlight the light collection capability of the detector and its ability to separate Cherenkov and scintillation hits in time.

Figure 9.9 shows a breakdown of the signal collection statistics for $0\nu\beta\beta$ events in the central 1 m of the detector; the results are summarised in Table 9.1. On average, 3299 hits are collected in each event, equivalent to 1319 h/MeV. 9.5% of the p.e. are lost to multi-hits. A further 19.8% are lost at the front end, with charges too small to trigger the discriminators. Only a very small fraction of hits are not collected by the trigger gate.

The light collection should be contrasted with the expected 350 h/MeV in SNO+ phase I for $r < 1$ m. The factor of 3.76 more hits can understood as:

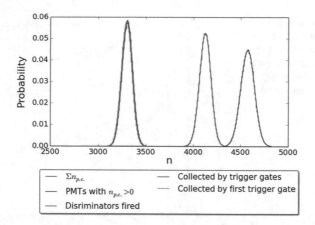

Fig. 9.9 Hit count for $0\nu\beta\beta$ events inside the central 1 m of the detector

Table 9.1 Hit breakdown for $0\nu\beta\beta$ events filling the central 1 m of the detector. No cuts applied

Description	Mean	Standard deviation
Photo-electrons	4563	71
PMTs with 1 or more p.e.	4126	61
Discriminators firing	3307	55
Collected by trigger gates	3299	55
Collected by 1st trigger gate	3299	55

$$\frac{1.37}{1} \times \frac{1.62}{1}$$

$$(\text{Effective Coverage}) \cdot (\text{PMT efficiency})$$

$$\times 1.6 \times \frac{1}{0.96} \times \frac{0.905}{0.91}$$

$$\times (\text{Light yield}) \times (\text{Secondary fluor efficiency}) \times (\text{Multihit effect})$$

$$= 3.68$$

the secondary fluor efficiency term arises from the re-emission probability of bisMSB in the SNO+ scintillator cocktail. 0.91 is the single p.e. fraction for $0\nu\beta\beta$ events with $r < 1$ m as measured in RAT 6.1.6 (Sect. 3.1.2).

Simply counting hits in this detector, one would achieve a 1.7% energy resolution, equivalent to 42.5 keV, for $0\nu\beta\beta$ at the centre, ignoring systematic effects. Note that the distributions are in general slightly broader than the Poisson limit $\sigma = \sqrt{N_{hit}}$. The additional width is likely due to variation in solid angle and attenuation as a function of position, which, to some extent, can be corrected for with knowledge of the vertex position.

Fig. 9.10 Photo-electron detection times for 2.5 MeV electrons in the central 1 m of the detector. The bumps in the region 150–200 ns are multiple reflections from the PMTs and AV

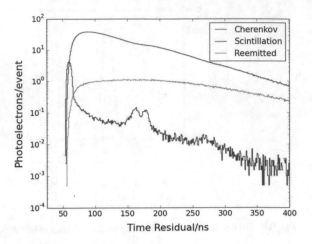

Figure 9.10 shows the creation time of p.e. produced by the average 2.5 MeV electron event that occurs the central 1 m of the detector. Even without correcting for time of flight, there is clear timing separation between the Cherenkov, scintillation and re-emission components.

9.2 Reconstruction

Cherenkov photons are emitted before scintillation photons, but this difference is only useful if the *detection* time can be related to the *emission* time by correcting for time of flight, which requires reconstruction of the event vertex.

Although an overall advantage, sensitivity to Cherenkov light complicates reconstruction, relative to SNO+, in three ways. First, the event position, time and direction(s) are all degenerate, so all three must be estimated together. Second, the light emitted is no longer isotropic and the angular distribution depends on the emission time. This necessitates 2D PDFs in time residual and detector angle. Third, the inferred vertex depends strongly on the event type because the amount and distribution of Cherenkov light depends on the number, energy and type of particles emitted.

This section describes reconstruction algorithms used to reconstruct the position, time and direction of point-like events using the maximum likelihood method.

9.2.1 Likelihood

The goal of the maximum likelihood method is to find the values of several fit parameters that maximise the probability of the observation made. Each event contains a set of time, position pairs $\{\vec{x}_{hit}, t_{hit}\}$, with one entry for each hit. For point-like events, the time of the event t_v and its position $\vec{x}_v$ must be estimated, along with a set of other parameters $\{\alpha\}$ that depend on the specific hypothesis considered.[3] The likelihood is therefore:

$$\mathcal{L} = P(\{\vec{x}_{hit}^i, t_{hit}^i\}|\vec{x}_v, t_v, \{\alpha\}) \tag{9.2}$$

The probability in Eq. 9.2 contains an enormous amount of detail. In principle, P depends on the physics that produced the primary particle, the emission and propagation physics of optical photons and the response of every PMT. To break this down into something more tractable, the first key assumption is that the hits are independent[4]:

$$\mathcal{L} = \Pi_{i=0}^{N_{hits}} P(\vec{x}_{hit}^i, t_{hit}^i|\vec{x}_v, t_v, \{\alpha\}) \tag{9.3}$$

Next it is assumed that hits are dominated by photons that followed straight line paths. Then, the propagation time of each photon can be corrected for using time residuals:

$$t_{res}^i = t_{hit} - t_v - t_{t.o.f} \tag{9.4}$$

Rewriting the likelihood in terms of time residuals absorbs the dominant dependence on $\vec{x}_v, t_v$:

$$\mathcal{L} = \Pi_{i=0}^{N_{hits}} P(t_{res}^i(\vec{x}_v, t_v), \vec{x}_{hit}^i|\vec{x}_v, t_v, \{\alpha\}) \tag{9.5}$$

One factor that is useful to isolate is the solid angle of the PMT as viewed from the vertex position. For the same emission time and direction, PMTs further away or at more oblique angles are less likely to be hit. With this correction the likelihood reads:

$$\mathcal{L} = \Pi_{i=0}^{N_{hits}} \frac{\Omega^i}{4\pi} P(t_{res}^i(\vec{x}_v, t_v), \vec{x}_{hit}^i|\vec{x}_v, t_v, \{\alpha\}) \tag{9.6}$$

where

$$\Omega^i = \vec{x_{pmt}} \cdot \frac{\vec{x_{pmt}} - \vec{x}_v}{\|\vec{x_{pmt}}\| \|\vec{x_{pmt}} - \vec{x}_v\|^3} \tag{9.7}$$

A similar term can be included to represent the optical absorption of LABSS: photons taking longer paths through the scintillator are attenuated more than those taking short paths. With this term, the final likelihood is:

$$\mathcal{L} = \Pi_{i=0}^{N_{hits}} \frac{\Omega^i}{4\pi} P(t_{res}^i(\vec{x}_v, t_v), \vec{x}_{hit}^i|\vec{x}_v, t_v, \{\alpha\}) \exp\left(\frac{-\Delta l}{l_{abs}}\right) \tag{9.8}$$

---

[3]E.g. for reconstructing electron events we will want to estimate the electron direction so $\{\alpha\} = \{\theta_e, \phi_e\}$.

[4]The photons propagate independently and they are emitted independently provided the hypothesis considered is approximately simple.

where Δl is the photon's path length in scintillator and l_{abs} is the total absorption length of LABSS at a representative wavelength.

The justification for explicitly including these terms is explored in detail in Appendix B. With them explicitly factored out, $P(t^i_{res}, \vec{x}^i_{hit}|\vec{x}_v, t_v, \{\alpha\})$ is the probability of observing a hit at position $\vec{x}_{pmt}$ with time residual t^i_{res}, corrected for solid angle and attenuation effects. It is dominated by variations in photon emission, but it also includes the PMT transit time smearing and the effect of non-straight line paths. Critically, if a symmetry can be exploited to reduce its dimensionality further, it can be estimated using a large number Monte Carlo events for which the truth values of $\vec{x}_v$, t_v and $\{\alpha\}$ are known.

The choice of attenuation length l_{abs} and the light velocity used to calculate $t_{t.o.f}$ is non-trivial because both are wavelength dependent. Hand tuning for optimal reconstructed resolution on 2.5 MeV electron events gives:

$$\lambda_{abs} = 50 \, \text{m} \qquad c_{eff} = 197 \, \text{mm ns}^{-1} \tag{9.9}$$

this procedure is described in Appendix C.

9.2.2   Optimisation

Point estimates for $\vec{x}_v$, t_v, $\{\alpha\}$ were calculated by maximising the likelihood in Eq. 9.8 with respect to those $4 + |\{\alpha\}|$ parameters. In practice, it is numerically expedient to maximise the natural logarithm of the likelihood. Dropping insignificant constant factors, the log-likelihood is:

$$\log \mathcal{L} = \sum_0^{N_{hit}} \left(\log P(t^i_{res}(\vec{x}_v, t_v), \vec{x}^i_{hit}|\vec{x}_v, t_v, \{\alpha\}) + \log \Omega^i - \frac{\Delta l}{l_{abs}} \right) \tag{9.10}$$

A simplex simulated annealing algorithm from numerical recipes [12] was used to maximise Eq. 9.10 with respect to the Cartesian coordinates of the vertex position x_v, y_v, z_v, the vertex time t_v and the hypothesis dependent $\{\alpha\}$. For each event the annealing algorithm was run 4 times, and the fit with the greatest likelihood was selected.

The optimisation routine is seeded with estimates $\{x_s, y_s, z_s, t_s\}$ and errors $\{\sigma_{x_s}...\}$ for the position and time from the standard ScintFitter described in Chap. 3. The PDF and effective speed for this fitter were tuned for the LABSS cocktail by E. Leming.

9.2.3 Electron Reconstruction

Rejecting the ^{8}B ν ES background requires reconstruction of electrons at around 2.5 MeV. For such events, the direction of the scattered electron is defined by $\hat{d}_v = (1, \theta_v, \phi_v)$, in detector coordinates.

$$P(t_{res}^i, \vec{x}_{hit}^i | \vec{x}_v, t_v, \{\alpha\}) = P(t_{res}^i, \vec{x}_{hit}^i | \vec{x}_v, t_v, \theta_v, \phi_v) \tag{9.11}$$

Note that the the explicit dependence of t_{res}^i on the vertex time and position has been dropped for brevity. By symmetry, the directional variation in probability should depend primarily on the angle, θ, made between the electron direction and the line pointing from the vertex to the hit PMT:

$$\cos \theta^i = \hat{d}_v \cdot \frac{\vec{x_{pmt}} - \vec{x}_v}{\|\vec{x_{pmt}} - \vec{x}_v\|} \tag{9.12}$$

then

$$P(t_{res}^i, \vec{x}_{hit}^i | \vec{x}_v, t_v, \theta_v, \phi_v) = P(t_{res}^i, \cos \theta^i) \tag{9.13}$$

Now the hit probability is a 2D distribution of variables t_{res}^i, $\cos \theta^i$ that depend only on the parameters of interest: $\vec{x}_t, t_v, \theta_v, \phi_v$.

A binned estimate of the PDF $P(t_{res}^i, \cos \theta^i)$ was calculated using 1245000 2.5 MeV Monte Carlo electron events, generated in the central 1 m of the detector. For each hit in the first triggered event of each simulated electron, the time residual was calculated according to Eq. 9.4, the angle with respect to the electron direction was calculated according to Eq. 9.12 and these values were filled into a 2D histogram. 2D visualisations of the PDF are shown in Fig. 9.11 and its 1D projections are shown in Fig. 9.12. The cut off at around 500 ns is the edge of the trigger gate; noise hits can sit outside this window after time of flight correction. Events falling after 540 ns were assigned a probability at the noise level of 10^{-8}, to which no solid angle or attenuation length correction was made. The binned PDF was splined in both directions using linear interpolation to prevent discontinuities in the PDF that can cause problems in optimisation.

In addition to the position and time seed described in Sect. 9.2.2, a direction seed $\{\theta_s, \phi_s\}$ was calculated by taking the average of the unit vectors that point from the seed position to each hit (Eq. 9.14). Only hits within a coarse Cherenkov time cut of $55 < t_{hit}/\text{ns} < 65$ were included. A fixed error of 0.5 was assumed for σ_ϕ, σ_θ.

Table 9.2 shows the full annealing parameter set used to fit events under the electron hypothesis.

$$\hat{d}_s = \frac{\vec{d_{seed}}}{\|\vec{d}_s\|} \qquad d_s = \sum_{i=0}^{N_{early}} \frac{\vec{x}_{pmt}^i - \vec{x}_s}{\|\vec{x}_{pmt}^i - \vec{x}_s\|} \tag{9.14}$$

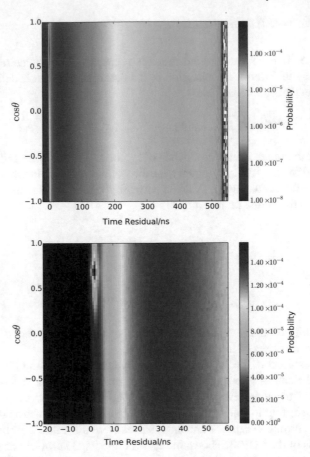

Fig. 9.11 2D PDF in time residual and $\cos\theta$ for 2.5 MeV electron events within the central 1 m of the detector, shown on two time scales. Bin widths are 770ps in time residual and 0.04 in $\cos\theta$

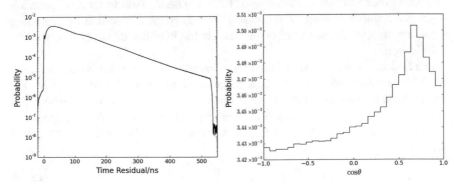

Fig. 9.12 1D PDF projections of the 2D PDF used to fit electron events

Table 9.2 Simulated annealing parameters for the electron fit

Parameter	Value
Iterations	100
Tolerance	10^{-4}
Schedule steps	10
Schedule power	4
x_{init}	x_s
y_{init}	y_s
z_{init}	z_s
σ_x	σ_{x_s}
σ_y	σ_{y_s}
σ_z	σ_{z_s}
θ_{init}	θ_s
ϕ_{init}	ϕ_s
σ_θ	0.5
σ_ϕ	0.5
t_{init}	t_{seed}
σ_t	σ_{t_s}

9.2.3.1 Performance

Figure 9.13 shows the performance of the fitter in reconstructing the time, position and direction of 2.5 MeV electrons. Table 9.3 summaries these plots using fits to functional forms. The error on the reconstructed position is mostly Gaussian distributed, but there is a clear non-Gaussian tail. The resolution in x and y is 11 cm whereas the resolution in the z direction is closer to 10 cm. 85% of events reconstruct to within 60° of the true direction.

There are three measurable biases. The first is a shift along the direction of motion of the electron, and a shift to later times. Shown in Fig. 9.13e, the drive is defined as the displacement from the truth to the reconstructed vertex, projected onto the true electron direction:

$$\text{Drive} = (\vec{x}_{fit} - \vec{x}_{truth}) \cdot \hat{d}_v \qquad (9.15)$$

On average, the algorithm reconstructs events with positive drive. The effect is strongly correlated with over-estimates of the event time and to poor estimates of the electron direction, as shown in Fig. 9.14. The effect can be understood as follows: multiple scattering of the primary electron and Rayleigh scattering of the photons it produces tends to push Cherenkov hits to angles *outside* the Cherenkov cone, whereas scatters into the cone are less likely. On average, this widens the emission cone. Pulling along the direction of motion makes this cone seem the correct width again, and the time of the event can be moved forward to compensate.

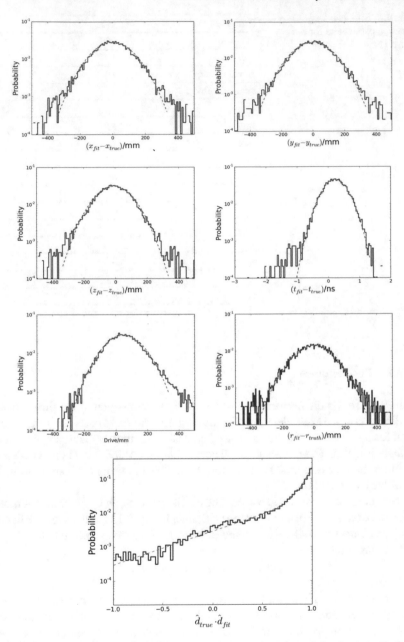

Fig. 9.13 Fit performance for 2.5 MeV electrons in the central 1 m of the detector. Each plot contains 9647 entries. 'Drive' shows the position resolution along the direction of the true electron momentum

Table 9.3 Electron reconstruction summary. Each parameter was estimated using a sample of 9647 events

Distribution	Unit	Model	Fit parameters
$x_{fit} - x_{truth}$	mm	Gaussian	$\mu = 1.74 \pm 0.02, \sigma = 114$
$y_{fit} - y_{truth}$	mm	Gaussian	$\mu = -0.29 \pm 0.03, \sigma = 115$
$z_{fit} - z_{truth}$	mm	Gaussian	$\mu = -10.1 \pm 0.1, \sigma = 102$
$r_{fit} - r_{truth}$	mm	Gaussian	$\mu = 0.304 \pm 0.003, \sigma = 116$
Drive	mm	Gaussian	$\mu = 49.2 \pm 0.5, \sigma = 109$
$t_{fit} - t_{truth}$	ns	Gaussian	$\mu = 0.219 \pm 0.002, \sigma = 0.3$
$\hat{d}_{true} \cdot \hat{d}_{fit}$	–	Double exponential	$N_1 = 2.86 \times 10^{-3}$, $N_2 = 8.17 \times 10^{-8}$
			$\tau_1 = 0.44, \tau_2 = 0.068$

The second bias is a pull to lower z-values. This is caused by a tendency for the fitter to choose directions close to the poles at $\theta = 0, \pi$, as shown in Fig. 9.15. This is likely caused by a slight asymmetry in the PMT placement (Fig. 9.8) that produces a local minimum there. This is also the cause of the smaller resolution in the z direction.

Finally, on average, $\cos\theta < 1$. This effect is caused by the hard boundary at $\cos\theta = 1$, which introduces a bias into the maximum likelihood method itself.

9.2.3.2 Sources of Error

Figure 9.16 shows the CDF for the $\cos\theta$ distribution in Fig. 9.13 including a breakdown of the effects that contribute to its width. The plot shows that the majority of the width comes from electron multiple scattering (an irreducible effect) and misreconstruction of the vertex position. On the other hand, background scintillation hits have next to no effect.

Slowing down the the scintillation light reduces the scintillation background inside the Cherenkov window, but it also hurts the detector's position resolution, which relies on the prompt peak. The scintillator considered here is in the extreme regime where the scintillation background is irrelevant, and the direction uncertainty is dominated by vertex error. This suggests the interesting possibility that a somewhat faster scintillator could perform better.

9.2.4 $0\nu\beta\beta$ Reconstruction

Reconstructing the time and position of electrons required the estimation of two internal degrees of freedom for each event: the θ, ϕ coordinates of the electron direc-

Fig. 9.14 The drive reconstruction bias for 2.5 MeV electrons. Positive drive values are correlated with late time estimates ($t_{fit} - t_{true} > 0$), and poor estimates of the electron direction ($\hat{d}_{fit} \cdot \hat{d}_{true} < 0$)

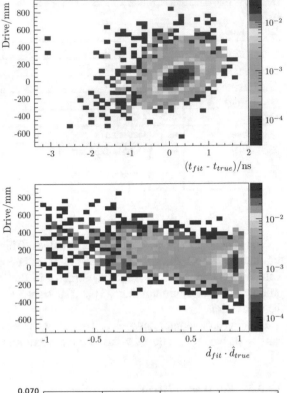

Fig. 9.15 Pull of the fit direction towards the poles for 2.5 MeV electrons. θ is the polar angle about the z-axis

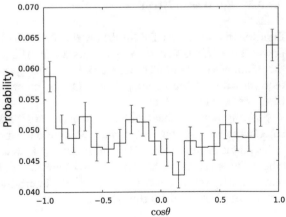

tion. Unfortunately, $0\nu\beta\beta$ events are much more complex. There are two electrons, emitted in two directions $\hat{d}_1$, $\hat{d}_2$ with a variable split of energy between them. In total there are $4 + 1 = 5$ internal degrees of freedom. Worse still, the energy split and angular correlation of the electrons are strongly model dependent (Sect. 1.7.4). This section describes three models of $0\nu\beta\beta$ of increasing complexity used to reconstruct

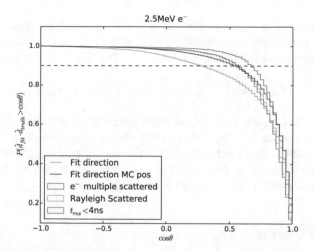

Fig. 9.16 Breakdown of effects creating width in the reconstructed direction. The 'e⁻ multiple scattered' line shows the distribution in the angle between the true electron direction, and the average emission direction of the Cherenkov photons it produces. The 'Rayleigh Scattered' line shows the mean direction of the Cherenkov hits w.r.t to the true vertex position and the '$t_{res} < 4$ ns' curve shows the mean direction of all hits in the first 4 ns (i.e. including background scintillation hits). Also shown is the performance of the fitter when the true vertex position is known exactly: 'Fit direction MC pos'

$0\nu\beta\beta$ events. The first two models assume the light neutrino exchange mechanism (LNE), whereas the third is model independent.

9.2.4.1 Isotropic Model

To zeroth order, the Cherenkov light produced in $0\nu\beta\beta$ events is isotropic. For LNE events, the electrons tend to separate back-to-back, both undergo multiple scattering and the photons they produce are scatted, absorbed and re-emitted. Thus, the simplest treatment of these events is to ignore angular information altogether. Then the hit probability is:

$$P(t_{res}^i, \vec{x}_{hit}^i | \vec{x}_v, t_v, \vec{\alpha}) = P(t_{res}^i) \tag{9.16}$$

The corresponding PDF can be built in the same way as the electron PDF (Sect. 9.2.3) except calculating $\cos\theta$ for each hit using a random direction. The simulated annealing parameters used to optimise this likelihood are the same as for electrons.

9.2.4.2 Unidirectional Model

Of course, any given event does actually have two associated directions, so a first order extension is to fit for the direction of the highest energy electron and to treat the contribution from the second electron as isotropic. In this description, $0\nu\beta\beta$ events look similar to electron events but with less light in the Cherenkov cone and an extra isotropic contribution from the second electron. The hit probability is:

$$P(t^i_{res}, \vec{x}^i_{hit}|\vec{x}_v, t_v, \vec{\alpha}) = P(t^i_{res}, \vec{x}^i_{hit}|\vec{x}_v, t_v, \theta_v, \phi_v) \qquad (9.17)$$

where θ_v, ϕ_v are the polar coordinates of the direction of most energetic electron.

The corresponding PDF can be built using the procedure described in Sect. 9.2.3, calculating $\cos\theta^i$ with respect to the initial direction of the highest energy electron.

The simulated annealing parameters used to optimise this likelihood are the same as for electrons and the direction seed is the same.

9.2.4.3 Bidirectional Model

The third and most complex hypothesis considered here includes the direction of both electrons. In this case, the $0\nu\beta\beta$ event is described by two electrons, emitted in two different directions but at the same time and position. The hit probability is:

$$\begin{aligned} P(t^i_{res}, \vec{x}^i_{hit}|\vec{x}_v, t_v, \vec{\alpha}) = f \cdot P(t^i_{res}, \vec{x}^i_{hit}|\vec{x}_v, t_v, \theta_1, \phi_1) \\ +(1-f) \cdot P(t^i_{res}, \vec{x}^i_{hit}|\vec{x}_v, t_v, \theta_2, \phi_2) \end{aligned} \qquad (9.18)$$

where f is a weighting between the contribution of the two electrons, loosely related to the energy divide between them. $\theta_1, \phi_1, \theta_2, \phi_2$ are the spherical coordinates of the two directions.

Equation 9.18 contains two copies of the following probability distribution, one for each electron:

$$P(t^i_{res}, \vec{x}^i_{hit}|\vec{x}_v, t_v, \theta, \phi) \qquad (9.19)$$

which is the probability of observing a hit at $\vec{x}^i$ with residual t^i_{res} after a single electron event at $\vec{x}_v, t_v$ with direction θ, ϕ. The energy of the electron events used to build this PDF sets the overall Cherenkov to scintillation ratio of the model; 1.25 MeV was chosen here because this is the expectation value of the electron energies under LNE. The PDF in Eq. 9.19 was built in exactly the same way as the single electron PDF, but using 1.25 MeV electrons.

f has hard boundaries at 0 and 1 which can cause problems for optimisation algorithms. To remedy this, f is transformed to a cyclic parameter χ defined by:

$$\sin^2\chi = f \qquad \cos^2\chi = 1 - f \qquad (9.20)$$

In total, there are $3 + 1 + 4 + 1 = 9$ free parameters in the bidirectional fit. Table 9.4 summarises their annealing parameters. The direction of electron 1 (arbitrarily) was seeded using using the procedure described in Sect. 9.2.3. The second direction was seeded randomly. θ, ϕ have a fixed errors of $\pi/2$ and π respectively for

Table 9.4 Simulated annealing parameters for the bidirectional fit

Parameter	Value
Iterations	100
Tolerance	1×10^{-4}
Schedule steps	10
Schedule power	4
x_{init}	x_s
y_{init}	y_s
z_{init}	z_s
σ_x	σ_{x_s}
σ_y	σ_{y_s}
σ_z	σ_{z_s}
$\theta_{1_{init}}$	θ_s
$\phi_{1_{init}}$	ϕ_s
$\theta_{2_{init}}$	$\sim \mathcal{U}(0, \pi)$
$\phi_{2_{init}}$	$\sim \mathcal{U}(-\pi, \pi)$
σ_{θ_1}	$\pi/2$
σ_{ϕ_1}	π
σ_{θ_2}	$\pi/2$
σ_{ϕ_2}	π
χ_{init}	$\sim \mathcal{U}(0, \pi/2)$
σ_χ	0.07
t_{init}	t_{seed}
σ_t	σ_{t_s}

both electrons. The χ seed is a random number between 0 and $\pi/2$ with an assumed error of 0.07.

PDFs for these three hypotheses are shown in Fig. 9.17, all three were splined in the fit using linear interpolation.

9.2.4.4 Performance

Figure 9.18 shows the x, y, z, r, t distributions for $0\nu\beta\beta$ events reconstructed under each of the three hypotheses described above. The drive distributions with respect to both electrons are also shown. Figure 9.19 shows the performance of the direction reconstruction for both electrons. Table 9.5 shows fit parameters derived from these distributions.

The fit resolution in x, y, z, r and t improves with model complexity. All three models exhibit the drive effect discussed in Sect. 9.2.3.1 for both electron directions, but the size of the effect is reduced with increased model complexity. The bidirectional fit has x and y resolutions of 13 cm, a z resolution of 12 cm and an average drive of 2 cm along the direction of both electrons.

Fig. 9.17 $0\nu\beta\beta$ PDFs.
Top to bottom: isotropic,
unidirectional and
bidirectional models

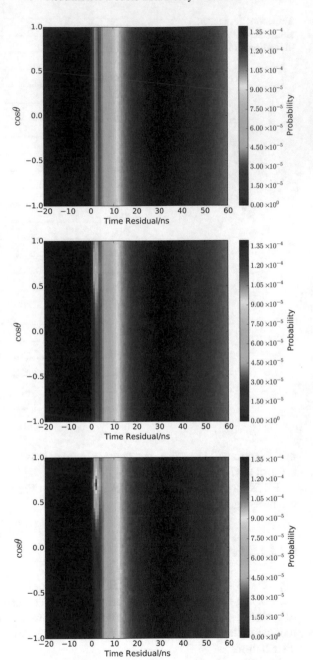

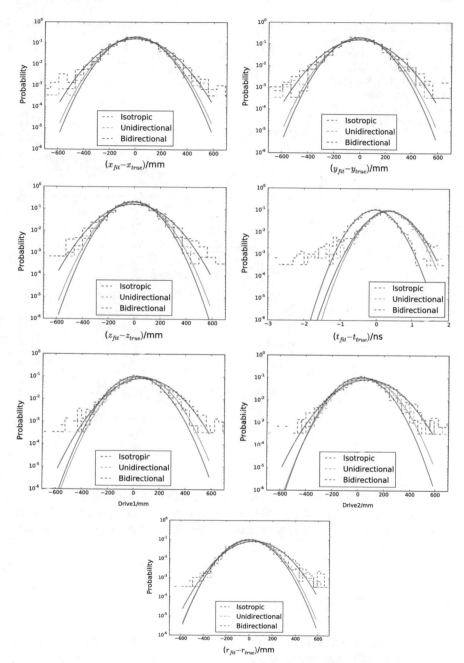

Fig. 9.18 Position fit performance for $0\nu\beta\beta$ in the central 1 m of the detector. Each plot contains 2924 entries. Drive1 and Drive2 show the resolution along in the direction of the momenta of electrons 1 and 2

Fig. 9.19 Direction fit
performance for $0\nu\beta\beta$ in the
central 1 m of the detector.
Each plot contains 2924
entries. The (1, 2) label is
meaningful because
electron1 is associated with
the seed direction

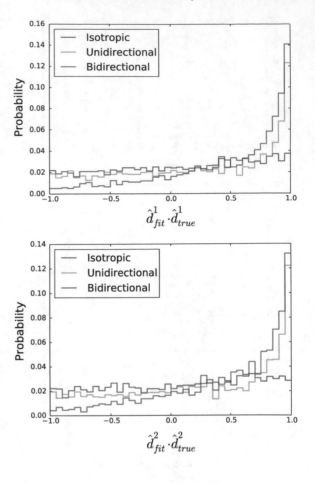

Electron 1 is, on average, fit better than electron 2 because direction 1 is (arbitrarily) the one that receives the calculated seed. The drive is larger for electron 1 because the seed will tend to point towards the electron that produces most Cherenkov light.

9.3 ^{8}B ν ES Rejection

With the ability to reconstruct event vertices, one can exploit differences in the Cherenkov signals produced by $0\nu\beta\beta$ and ^{8}B ν ES events. There are three: first, for events of the same overall energy, there are more Cherenkov photons for single electrons than for two electron $0\nu\beta\beta$ events, therefore, the ^{8}B ν ES events should have a larger ratio of Cherenkov hits to scintillation hits; second, the Cherenkov light is more isotropic for $0\nu\beta\beta$ events than electron events, because the two electrons are

Table 9.5 $0\nu\beta\beta$ fit summary, each parameter was estimated using a sample of 2924 events

Distribution	Unit	Model	Parameters	Isotropic	Unidirectional	Bidirectional
$x_{fit} - x_{truth}$	mm	Gaussian	μ	0.76 ± 0.01	0.08 ± 0.002	1.21 ± 0.02
			σ	157.2	135.3	128.5
$y_{fit} - y_{truth}$	mm	Gaussian	μ	-0.67 ± 0.01	0.98 ± 0.02	-2.8 ± 0.1
			σ	160.1	134.7	126.4
$z_{fit} - z_{truth}$	mm	Gaussian	μ	-6.0 ± 0.1	-6.3 ± 0.1	-8.8 ± 0.2
			σ	154.7	127.0	118.0
$r_{fit} - r_{truth}$	mm	Gaussian	μ	36.9 ± 0.7	7.0 ± 0.1	-4.5 ± 0.08
			σ	153.8	132.3	128.2
$t_{fit} - t_{truth}$	ns	Gaussian	μ	0.35 ± 0.01	0.33 ± 0.01	-0.046 ± 0.001
			σ	0.41	0.39	0.37
Drive1	mm	Gaussian	μ	71.8 ± 1.3	54.9 ± 1.0	21.9 ± 0.4
			σ	152.8	127.6	124.1
Drive2	mm	Gaussian	μ	69.3 ± 1.2	49.0 ± 0.9	20.1 ± 0.4
			σ	154.5	128.2	120.0

emitted in different directions; finally, the direction of Cherenkov photons produced in ^{8}B ν ES events is strongly correlated with the direction to the sun, whereas $0\nu\beta\beta$ events have no correlation with the sun's position.

The first half of this section discusses each of these handles in turn, devises simple cuts and combines them. The second half applies a likelihood-ratio method that makes use of the same information indirectly. The two approaches are compared and an estimate of the sensitivity improvement for SNO+ style $0\nu\beta\beta$ searches is calculated.

Throughout this chapter, events referred to as ^{8}B only include those events depositing 2.4–2.6 MeV in the scintillator, because these are the events that produce a $0\nu\beta\beta$ background.

9.3.1 Solar Direction

The electrons produced in ^{8}B events tend to point away from the sun, because the neutrinos travel in straight lines, and because the neutrino-electron interaction is forward pointing. However, the reconstructed solar angle, $\cos\theta_\odot$, of these events will have a finite width, because not all scatters are forward, the electron multiple scatters, and because of the finite direction resolution of the detector.

The first factor is particularly important at 2.5 MeV. Figure 9.20 shows the distribution of the angle made between ^{8}B solar neutrinos and the electrons they produce in elastic scattering events. Counter-intuitively, it shows that the electron scatters at 25° from the neutrino on average. This can be understood from the kinematics of the neutrino-electron interaction. 2.5 MeV electrons may be produced by neutrinos with

Fig. 9.20 ^{8}B ν-e scattering
angles. θ is the smallest
angle that separates the
incoming neutrino direction
and the out-going electron
direction

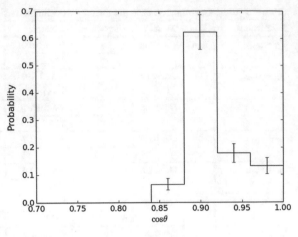

Fig. 9.21 Reconstructed
solar angles for $0\nu\beta\beta$ and ^{8}B
ν ES events. θ is the angle
between the fit direction and
the true direction, $\hat{d}_e$, or the
solar direction, $\hat{d}_\odot$. $r < 1$ m

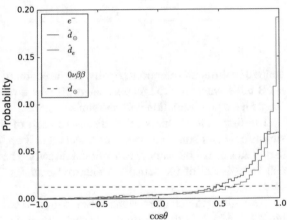

close to 2.5 MeV in energy, producing a co-linear electron, or by a higher energy
neutrino passing on a small fraction of its energy in a wide angle scatter. Because
most of the ^{8}B flux is above 2.5 MeV, the latter type of interaction is more common.

Figure 9.21 shows $\cos\theta_\odot$ for ^{8}B and $0\nu\beta\beta$ events alongside the fit resolution,
$\hat{d}_{fit} \cdot \hat{d}_{true}$ for ^{8}B events; it shows that neutrino-electron scattering adds significant
width to the $\cos\theta_\odot$ distribution and that, as expected, there is no correlation between
the sun's direction and the reconstructed direction of $0\nu\beta\beta$ events. To estimate the
power of these distributions, an arbitrary cut of $\cos\theta_\odot < 0.2$ was applied; it rejects
90% of ^{8}B events with a $0\nu\beta\beta$ sacrifice of 43%.

9.3.2 Early Hit Fraction

Figure 2.8 in Chap. 2 showed that the number of Cherenkov photons emitted by an electron is approximately linear with energy, above a threshold of 0.39 MeV. From this plot, two 1.25 MeV electrons should create approximately 80% of the Cherenkov photons produced by a single electrons. This difference should be reflected in the number of hits with early time residuals.

Figure 9.22 shows the time residuals for ⁸B ν ES and $0\nu\beta\beta$ events with respect to a vertex reconstructed *under the hypothesis that both are single electrons*. There is a small excess for electron events between time residuals of -1 and 3 ns.

An early hit fraction parameter was defined to exploit this difference. f_{early} is the the fraction of hits that have time residuals between -1 and 3 ns:

$$f_{early} = \frac{N_{hit}(-1\,\text{ns} < t_{res} < 3\,\text{ns})}{N_{hit}} \tag{9.21}$$

Figure 9.22 shows the distributions of f_{early} for $0\nu\beta\beta$ and ⁸B ν ES events in the central 1 m. On average, the early hit fraction is 20% larger for ⁸B ν ES but the breadth of the two distributions means that they overlap significantly. This parameter has only weak statistical power for discriminating between $0\nu\beta\beta$ and ⁸B ν ES events.

9.3.3 Isotropy

The final handle is the isotropy of the Cherenkov light. If reconstructed as electron, $0\nu\beta\beta$ events should have more hits outside the apparent Cherenkov cone than single electron events. To exploit this difference, events were reconstructed as electrons, the Cherenkov signal was isolated using a time cut of $-1 < t_{res}/\text{ns} < 3$, and the apparent emission angle for each hit $\cos\theta_{em}$ was calculated using

$$\cos\theta_{em} = \hat{d}_{fit} \cdot \frac{\vec{x}_{pmt} - \vec{x}_{fit}}{|\vec{x}_{pmt} - \vec{x}_{fit}|} \tag{9.22}$$

These emission angles were then used to calculate the forward fraction, $f_{forward}$, defined by:

$$f_{forward} = \frac{N_{hit}(-1 < t_{res}/\text{ns} < 3 \text{ and } \cos\theta_{em} > 0.54)}{N_{hit}(-1 < t_{res}/\text{ns} < 3)} \tag{9.23}$$

Figure 9.23 shows the distributions in $\cos\theta_{em}$ and $f_{forward}$ for ⁸B ν ES and $0\nu\beta\beta$ events; the $f_{forward}$ distributions have been fitted to Gaussian distributions. On average, $f_{forward}$ is larger for ⁸B ν ES events but the width of the two distributions is again broad. The two peaks are separated by approximately half of their width.

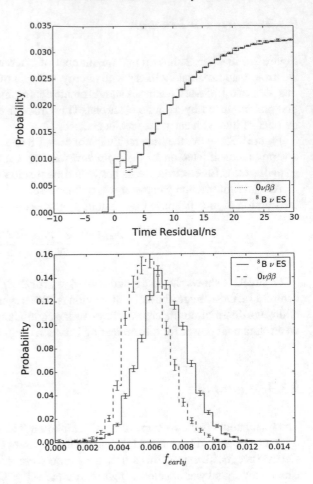

Fig. 9.22 Time residuals and early hit fraction for 2924 $0\nu\beta\beta$ events and 1938 ^{8}B ν ES events, $r < 1$ m. Error bars on the upper plot are too small to be seen. $r < 1$ m

9.3.4   Combining Discriminants

Figure 9.24 shows the correlations between the three discriminants described so far, $\cos\theta_\odot$, f_{early} and f_{front} for the two event classes. The correlations are weak, so there is merit to combining them into a single discriminant. Of the many ways of doing this, only a simple Fisher discriminant is investigated here, its distribution for both event types is shown in Fig. 9.25.

Figure 9.26 shows the background rejection as a function of the signal efficiency that can be achieved by cutting on each of the discriminants described above. Of the three primary cuts, the solar direction is by far the most powerful. The Fisher discriminant improves on the solar direction cut for $0\nu\beta\beta$ efficiencies >0.9 by including information from f_{early} and $f_{forward}$ (though the improvement is slight), but it does not add any power for smaller efficiencies. This can be understood from the sub-population of events that mimic the incorrect event class in each case. The $0\nu\beta\beta$

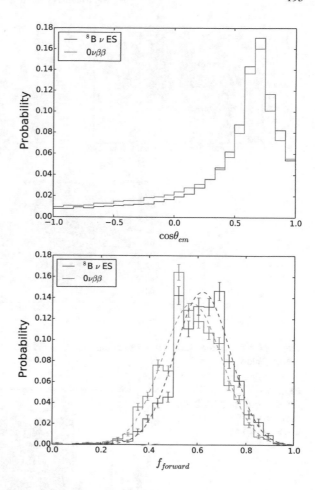

Fig. 9.23 Top: PDF of $\cos\theta_{em}$. Bottom: fraction of early hits with $\cos\theta_{em} > 0.54$ for 2924 $0\nu\beta\beta$ events and 1938 ^{8}B ν ES events, $r < 1\,\text{m}$

events which mimic ^{8}B are those that reconstruct along the solar direction *by chance*, so, on average, they still look different from ^{8}B in the other two discriminants. On the other hand, those ^{8}B events which mimic $0\nu\beta\beta$ will be those undergoing hard electron scatters or those reconstructed poorly; both cases make the events more $0\nu\beta\beta$-like in all three discriminants.

9.3.5 Likelihood-Ratio

An alternative approach is to reconstruct each event under both $0\nu\beta\beta$ and ^{8}B ν ES hypotheses and form a likelihood-ratio using the best fit likelihoods from each. The fit procedures described in Sects. 9.2.3 and 9.2.4 that fit for electrons and $0\nu\beta\beta$ events respectively produce an electron likelihood, $\mathcal{L}_{e^-}$, and a $0\nu\beta\beta$ likelihood, $\mathcal{L}_{0\nu}$. The

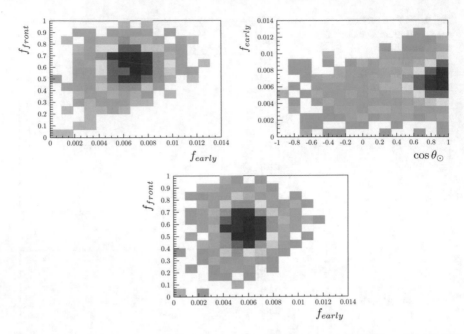

Fig. 9.24 Correlations between discriminants for ^{8}B ν ES background events (above) and $0\nu\beta\beta$ events (below). $r <1$ m

Fig. 9.25 Fisher discriminant for 2924 $0\nu\beta\beta$ events and 1938 ^{8}B ν ES events, $r < 1$ m

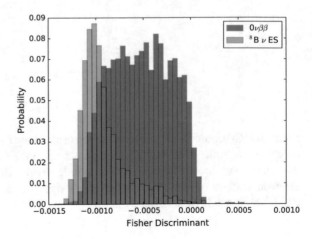

log-likelihood-ratio is:

$$\Delta \log \mathcal{L} = \log \left(\frac{\mathcal{L}_{0\nu}^{max}}{\mathcal{L}_{e^-}^{max}} \right) \tag{9.24}$$

This method comes with a couple of key advantages over the cut based approach. First, to apply the cuts, all events were reconstructed under a single reconstruction hypothesis, i.e. *both ^{8}B ν ES and $0\nu\beta\beta$ events were reconstructed as electrons*. This

Fig. 9.26 Cut efficiencies
for ^{8}B ν ES discrimination.
$r < 1\,\mathrm{m}$

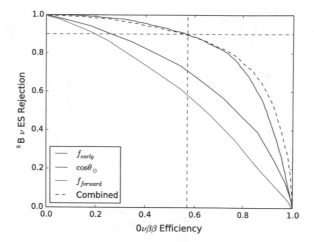

produces reconstruction biases that reduce the differences between the two event
classes, because the maximum likelihood fit chooses the vertex for which events best
match the target PDF. In other words, $0\nu\beta\beta$ events were reconstructed to *look most
electron-like*. Reconstructing twice instead allows each event a chance to match both
hypotheses, and the degree of agreement with each is estimated naturally from the
likelihood-ratio.

Second, if the PDF includes sufficient detail that the hypotheses are approximately simple, the Neyman–Pearson lemma guarantees that the likelihood will give
an optimal test statistic.

Figure 9.27 shows the log-likelihood-ratios for $0\nu\beta\beta$ and ^{8}B ν ES events for each
of the $0\nu\beta\beta$ hypotheses explored in Sect. 9.2.4. Positive values of $\Delta \log \mathcal{L}$ indicate
that the event looks more $0\nu\beta\beta$-like, whereas negative values indicate that event
appears more ^{8}B ν ES-like. Figure 9.28 shows the background rejections and signal
efficiencies that can be achieved cutting on these ratios. The unidirectional fit produces the best discrimination: 90% of ^{8}B ν ES events can be rejected with a 65%
sacrifice of $0\nu\beta\beta$. Note that this is significantly worse than the cut based approach,
because the likelihood does not yet include any information about the solar direction,
the most powerful handle for identifying these events.

9.3.6 Including Prior Information

The solar direction information can be naturally included in to the likelihood method
using a constraint on the reconstructed direction in the electron hypothesis, added as
a pre-factor to the likelihood in Eq. 9.8:

$$\mathcal{L}_e \rightarrow \mathcal{L}_e \cdot P(\cos \theta_\odot) \tag{9.25}$$

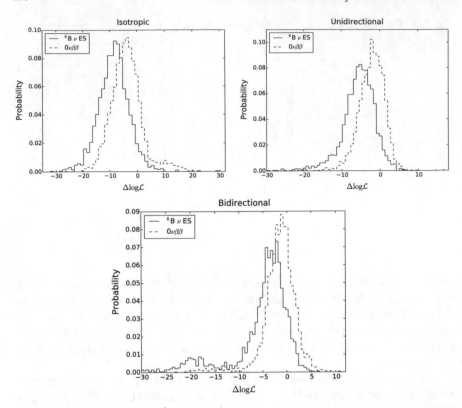

Fig. 9.27 $\Delta \log \mathcal{L}$ for $0\nu\beta\beta$ and $^8\text{B}\ \nu$ ES events. Positive $\Delta \log \mathcal{L}$ values indicate more $0\nu\beta\beta$-like events. $r < 1\,\text{m}$

Fig. 9.28 $^8\text{B}\ \nu$ ES rejection as a function of $0\nu\beta\beta$ efficiency for cuts on the likelihood-ratios in Fig. 9.27. $r < 1\,\text{m}$

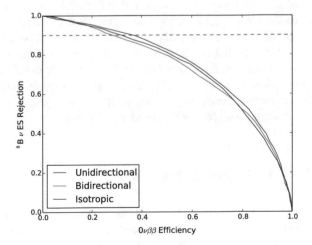

where $\cos \theta_\odot$ is the angle between the proposed electron direction $\hat{d}_e$ and the direction that points to the sun $\hat{d}_\odot$:

$$\cos \theta_\odot = \hat{d}_e \cdot \hat{d}_\odot \qquad (9.26)$$

$P(\cos \theta_\odot)$ is the probability of producing an electron at an angle of $\cos \theta_\odot$ in a ^{8}B ν ES event. At first glance Fig. 9.20 describes just that: the PDF of the cosine of the angle between the scattered electron direction and the parent neutrino, which points to the sun. However, electron multiple scattering means that an electron may emit many of its Cherenkov photons at large angles from its initial direction (Fig. 9.16); these events can have an apparent direction with $\cos \theta_\odot < 0.85$. If Fig. 9.20 was used as a constraint, it would assign these wide angles a probability of 0, which is clearly inconsistent. A better constraint should describe the *apparent* direction of the electron rather than the true direction. It should help to resolve the ambiguities created by vertex uncertainty, without over-correcting for effects that will always be apparent in the data.

The approach followed here was to estimate the PDF of the apparent solar angle in Monte Carlo. Section 9.2.3 showed that the performance of the electron fit on 2.5 MeV electrons is limited by uncertainty on the vertex position. To estimate what the angular distribution would look like without this uncertainty, a large number of ^{8}B ν ES events were fit as electrons, with the vertex time and position fixed at their truth values. Figure 9.29 shows the solar angles produced by this fit. The following function is a good fit to the distribution:

$$\begin{cases} a_1 \exp(\cos \theta_\odot / \tau_1) + a_2 \exp(\cos \theta_\odot / \tau_2) & \cos \theta_\odot < 0.87 \\ b & \cos \theta_\odot > 0.87 \end{cases} \qquad (9.27)$$

with $a_1 = 6.7 \cdot 10^{-6}$, $a_2 = 1.1 \cdot 10^{-3}$, $\tau_1 = 0.11$, $\tau_2 = 0.35$ and $b = 0.04$. The true vertex for real events is, of course, unknown, but this distribution can be estimated in Monte Carlo and applied as a constraint in Eq. 9.25.

Figure 9.30 shows likelihood-ratios calculated for $0\nu\beta\beta$ and ^{8}B ν ES with this constraint applied. As expected, the solar constraint drastically improves the separation of the two distributions. For the likelihood-ratio without the solar constraint, the $0\nu\beta\beta$ events that look most electron-like are those that look, by chance, most directional, but, more often than not, their apparent direction is not in the direction of the sun. The solar constraint uses this to better distinguish the two event types.

Figure 9.31 shows the background rejection factors and signal efficiencies that may be achieved using the likelihood-ratios with the solar constraint. The unidirectional and bidirectional fits both out-perform the combined cuts method. A cut used to eliminate 90% of ^{8}B ν ES events has a 5% greater signal efficiency for the likelihood methods than the combined cuts method.

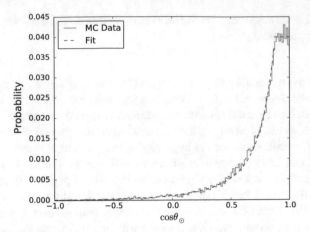

Fig. 9.29 PDF of $\cos\theta_\odot$ where $\theta_\odot$ is the smallest angle between the fit direction and the solar direction when the vertex position is fixed to its true value. The fit shown is used as a constraint on $\theta_\odot$ for the likelihood-ratio discriminant

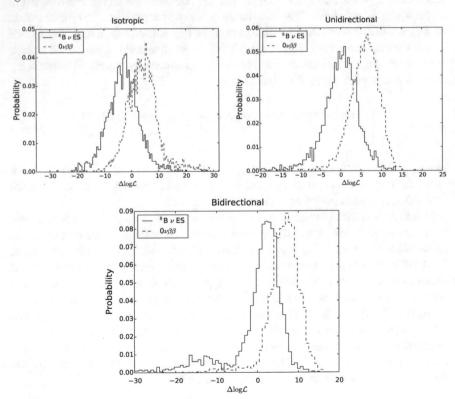

Fig. 9.30 Likelihood-ratios for $0\nu\beta\beta$ and ^{8}B ν ES events for the three $0\nu\beta\beta$ hypotheses using the solar constraint. Positive values indicate more $0\nu\beta\beta$-like events. $r < 1\,\mathrm{m}$

Fig. 9.31 ^{8}B ν ES rejection
as a function of $0\nu\beta\beta$
efficiency for cuts on the
likelihood-ratios with solar
constraint. A straightforward
cut on $\cos\theta_\odot$ is also shown
for comparison

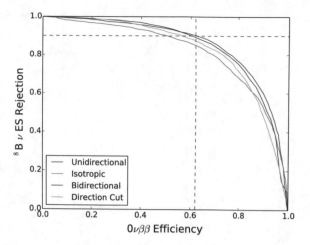

9.3.7 Significance

Assuming a $0\nu\beta\beta$ search is completely dominated by the ^{8}B ν ES background, the
sensitivity improvement from a ^{8}B cut is approximately[5]:

$$R = \frac{T_{1/2}^c}{T_{1/2}^u} = \frac{\epsilon_{\beta\beta}}{\sqrt{1 - r_\odot}} \tag{9.28}$$

where $T_{1/2}^c$, $T_{1/2}^u$ are the confidence limits on the $0\nu\beta\beta$ half-life with and without
the cut, $\epsilon_{\beta\beta}$ is the fraction of $0\nu\beta\beta$ events passing the cut and $r_\odot$ is the fraction of
solar neutrino events rejected by the cut. Figure 9.32 shows R as a function of $0\nu\beta\beta$
efficiency for the most effective cuts explored so far. It shows that the techniques
developed here could improve the sensitivity of a future $0\nu\beta\beta$ experiment, dominated
by ^{8}B ν ES, by a factor of 2.2 in $T_{1/2}$, or 1.5 in $m_{\beta\beta}$.

9.4 Determining the $0\nu\beta\beta$ mechanism

This final section discusses the possibility of distinguishing the underlying physics
mechanism of $0\nu\beta\beta$, using the light neutrino exchange (LNE) and right handed
current (RHC) mechanisms as examples. The two mechanisms differ in the angular
separation and energy split of the electrons they produce, shown again in Fig. 9.33. In
LNE, the two electrons tend to separate back to back with similar energies. In the RHC
mechanism, the electrons tend to be emitted in parallel, with the majority of the energy
going to a single electron. However, the two distributions overlap significantly, so *any*

---

[5]Assuming the number of background rate after cuts is Gaussian distributed.

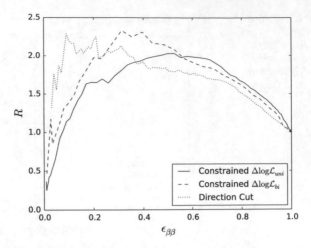

Fig. 9.32 Sensitivity improvements to the $0\nu\beta\beta$ half-life from applying cuts to $\cos\theta_\odot$ directly, or on likelihood ratios with the solar constraint

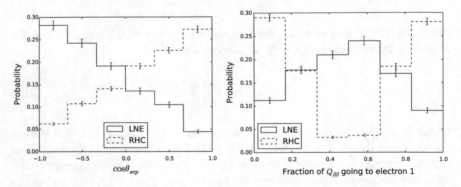

Fig. 9.33 Model dependence for $0\nu\beta\beta$ decay kinematics. Left: the cosine of the angle separating the emitted electrons (MC truth). Right: the fraction of the total energy release taken by electron 1 (arbitrarily chosen). Generated using `RAT 6.1.6`, which implements the distributions in [13]

test of the mechanism, using any technology, will rely on statistically distinguishing between them.

The bidirectional fit estimates the directions of both electrons $\hat{d}_{1,2}$. The apparent separation angle of the two electrons can be calculated using the dot product of these two directions:

$$\cos\theta_{sep} = \hat{d}_1 \cdot \hat{d}_2 \tag{9.29}$$

this should be sensitive to differences in the angular separation. The fit also estimates $\sin^2\chi$, which is equal to the conditional probability that any given photon came from electron 1 (roughly the fraction of Cherenkov photons produced by electron 1). This should be sensitive to the differences in the energy split.

Because $\cos\theta_{sep}$ is poorly constrained when $\sin^2\chi$ is close to 0 or 1, and $\sin^2\chi$ is poorly constrained close to $\cos\theta_{sep} = 1$, it is more useful to look at an apparent momentum imbalance. The two electrons have estimated kinetic energies of:

$$T_1 = \sin^2\chi \cdot Q_{\beta\beta} \tag{9.30}$$

$$T_2 = \cos^2\chi \cdot Q_{\beta\beta} \tag{9.31}$$

where $Q_{\beta\beta}$ is 2.5 MeV. This is related to the momentum by:

$$p = \sqrt{T(T + 2m_e)} \tag{9.32}$$

in natural units, where m_e is the electron mass. The momentum imbalance of the electron pair is $\mathcal{E}$, calculated as:

$$\mathcal{E} = |p_1\hat{d}_1 + p_2\hat{d}_2| \tag{9.33}$$

Figure 9.34 shows $\cos\theta_{sep}$, $\sin^2\chi$ and $\mathcal{E}$ for 0νββ events for the RHC and LNE mechanisms, 2.5 MeV e^- are also shown for reference. The fits on the left were performed by fixing the vertex position and time to their truth values, those on the right floated these four parameters.

The $\sin^2\chi$ and $\cos\theta$ look quite different to what one might naively expected from the theoretical distributions in Fig. 9.33. In particular, there is a strong tendency for both LNE and RHC to fit at $\sin^2\chi = 0, 1$ where one electron appears to take all the energy. This is the result of the Cherenkov threshold and fluctuations in the number of hits from each electron: both mechanisms sometimes emit electrons that produce few Cherenkov hits or are below threshold altogether.

Using the truth vertex, the $\mathcal{E}$ distributions are as expected: single electrons produce the most imbalanced $\mathcal{E}$, followed by RHC events followed by LNE. However, using the reconstructed position, the differences between the distributions are largely washed out.

Supposing a number of 0νββ events were observed, one could, for example, try to rule out the RHC mechanism in favour of LNE by comparing the observed $\mathcal{E}$ distribution against those in Fig. 9.34. If there are n_{obs} observed 0νββ events, with $\mathcal{E}$ values of $\{\mathcal{E}^i\}$, the log-likelihood-ratio comparing the LNE and RHC hypotheses is:

$$\mathcal{R} = \sum_{i=0}^{n_{obs}} \log\left(\frac{P_{LNE}(\mathcal{E}^i)}{P_{RHC}(\mathcal{E}^i)}\right) \tag{9.34}$$

Each hypothetical experiment will observe a different $\{\mathcal{E}^i\}$, so the distribution of $\mathcal{R}$ must be estimated with toy Monte Carlo. This was performed at a variety of n_{obs} for both mechanisms as follows:

1. Draw n_{obs} values from the LNE $\mathcal{E}$ distribution in Fig. 9.34.
2. Calculate the $\mathcal{R}$ of these observations according to Eq. 9.34.

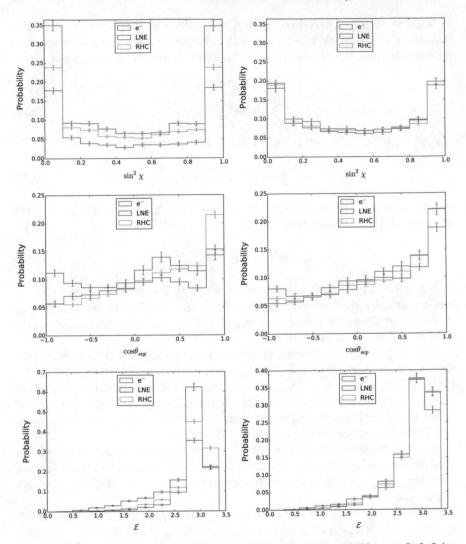

Fig. 9.34 $\sin^2 \chi$, $\cos \theta_{sep}$ and $\mathcal{E}$ for 2.5 MeV electrons, alongside LNE and RHC events. Left: fixing truth vertex time and position. Right: floating the vertex time and position. Calculated using 2924 LNE, 5123 RHC, and 1800 electrons

3. Fill a histogram with the $\mathcal{R}$ value.
4. Repeat steps 1–3 a large number of times.
5. Repeat steps 1–4 for the RHC mechanism.

Figure 9.35 shows these $\mathcal{R}$ distributions at a variety of n_{obs}. The reconstructed distribution is only a very weak discriminator of the mechanism, even if 50 events are observed.

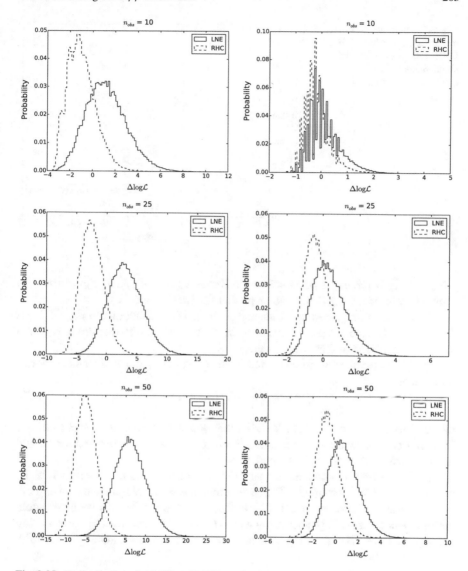

Fig. 9.35 $\mathcal{R}$ distributions for LNE and RHC mechanisms. Left: vertex time and position fixed to their truth values. Right: floating the vertex time and position. Each curve is built from 10^6 toy Monte Carlo experiments

The expected significance of any result can be estimated by calculating the p-value of the expected LNE signal under the RHC hypothesis. This is calculated according to:

$$p = P_{RHC}(\mathcal{R} < \overline{\mathcal{R}}_{LNE}|n_{obs}) \qquad (9.35)$$

Fig. 9.36 p-values for ruling
out the RHC mechanism if a
LNE bump is observed, in
standard deviation
equivalent, as a function
of n_{obs}

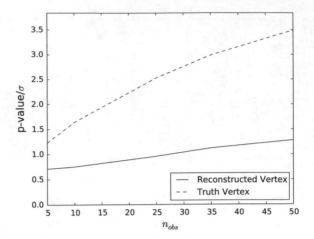

where $\overline{\mathcal{R}}_{LNE}$ is the expected value for LNE events given n_{obs} and P_{RHC} gives the
probability of observing that value or smaller for RHC events.

$\overline{\mathcal{R}}_{LNE}$ was estimated from the means of the LNE distributions in Fig. 9.35 and
P_{RHC} was estimated using the fraction of events in the RHC distributions in Fig. 9.35
that were smaller than that critical value.

Finally, the p-values were converted to an equivalent number of standard deviations using:

$$n_\sigma = \sqrt{2}\, \mathrm{Erf}^{-1}(2p - 1) \qquad (9.36)$$

the result is shown in Fig. 9.36. It shows that you would need 50 events to favour
LNE at 1σ. If instead the vertex position could be exactly determined, one could
reach 3σ with a pure signal of around 35 events.

This is the first demonstration that the $0\nu\beta\beta$ mechanism can, in principle, be
determined in liquid scintillator. The current model is severely limited by uncertainty
on the vertex position. Again, this hints that a somewhat faster scintillator could
perform better. A real implementation would also need to deal with a contamination
of single electron ^{8}B events; Fig. 9.34 demonstrates that this could be achieved with
the same $\mathcal{E}$ statistic.

References

1. Alonso JR et al (2014) Advanced scintillator detector concept (ASDC): a concept paper on the
   physics potential of water-based liquid scintillator. arXiv:1409.5864
2. Biller S (2017) Private communication
3. Wright A (2017) Te Dev R&D pre-summary. SNO+-docDB 4545-v1
4. Mastbaum A, Barros N, Coulter I, Kaptanoglu T, Segui L (2016) Optics overview and proposed
   changes to RAT. SNO+-docDB 3461

5. Kaptanoglu T (2017) Characterization of the Hamamatsu 8 R5912-MOD photomultiplier tube. arXiv:1710.03334
6. Klein J (2018) Private communication
7. Wang W et al (2015) Performance of the 8-in. R5912 photomultiplier tube with super bialkali photocathode. J Instrum 10(08):T08001. http://stacks.iop.org/1748-0221/10/i=08/a=T08001
8. Biller SD et al (1999) Measurements of photomultiplier single photon counting efficiency for the Sudbury Neutrino Observatory. Nucl Instrum Methods Phys Res Sect A: Accel Spectrom Detect Assoc Equip 432(2):364–373. https://doi.org/10.1016/S0168-9002(99)00500-8
9. Zhi Y, Liang Y, Wang Z, Chen S (2017) Wide field-of-view and high-efficiency light concentrator. arXiv:1703.07527
10. Boger J et al (2000) The Sudbury neutrino observatory. Nucl Instrum Meth 449:172–207
11. Akashi-Ronquest M et al (2015) Improving photoelectron counting and particle identification in scintillation detectors with Bayesian techniques. Astropart Phys 65:40–54. https://doi.org/10.1016/j.astropartphys.2014.12.006, arXiv:1408.1914
12. Press WH, Teukolsky SA, Vetterling WT, Flannery BP (2007) Numerical recipes 3rd edition: the art of scientific computing, 3rd edn. Cambridge University Press, New York
13. Tretyak VI, Zdesenko YuG (1995) Tables of double beta decay data. At Data Nucl Data Tables 61(1):43–90. https://doi.org/10.1016/S0092-640X(95)90011-X

Chapter 10
Conclusions

This thesis has explored the use of PMT hit patterns in time and space for reconstruction and particle identification in liquid scintillator, as well as the application of Bayesian methods to $0\nu\beta\beta$ signal extraction.

Using Hamiltonian Markov Chain Monte Carlo, two dimensional fits in event energy and radius were employed, to predict a SNO+ $0\nu\beta\beta$ half-life sensitivity of $T_{1/2}^{\beta\beta} > 1.76 \times 10^{26}$ yr, at 90% confidence, after a three year live-time.

$\beta^{\pm}\gamma$ events produced by radioactive decay in the scintillator were shown to have non-point-like timing distributions, produced by the multi-site deposition of Compton scattering γ and the time delay caused by ortho-positronium formation. This characteristic signature was used to differentiate internal backgrounds from point like $0\nu\beta\beta$ events using pulse shape discrimination (PSD) parameters. In particular, a PSD parameter, designed to separate $0\nu\beta\beta$ from poorly constrained ^{60}Co decay, was used to improve the 3σ $m_{\beta\beta}$ discovery level from 191 meV, which is already ruled out by Kamland-Zen, to 90.5 meV, which is allowed by all experiments.

Similarly, 40–60% rejection of each of the dominant external backgrounds inside $r < 4.2$ m was demonstrated using PSD, without significant sacrifice. For these backgrounds, improved rejection was achieved by accounting for the timing correlations and the angular hit distribution of the external backgrounds.

Finally, using a simulation of a next-generation slow-scintillator detector, equipped with a high coverage of high quantum efficiency, fast PMTs, it was shown that the angular distribution of Cherenkov light and the timing distribution of scintillation light can be used to reconstruct the position, time and direction of electrons. This information can be used to reject the ^{8}B elastic scattering background, improving the $m_{\beta\beta}$ sensitivity of a ^{8}B dominated experiment by 50%, and, in principle, to determine the underlying mechanism of $0\nu\beta\beta$.

© Springer Nature Switzerland AG 2019
J. Dunger, *Event Classification in Liquid Scintillator Using PMT Hit Patterns*,
Springer Theses, https://doi.org/10.1007/978-3-030-31616-7_10

Appendix A
Background Data

This chapter contains the supplementary data for Chap. 4, including the background rates assumed for the sensitivity studies in Chap. 8. Entries marked with a $*$ were included in the $0\nu\beta\beta$ fit (Tables A.1, A.2, A.3, A.4 and A.5). Figure A.1 shows the decay scheme for ^{208}Tl.

Table A.1 Expected cosmogenic decay rates for problem isotopes created in tellurium after one year's surface exposure [1]

Isotope	Events/t	
^{110}Ag	401.17	
$^{110(m)}$Ag	2.98×10^4	
^{60}Co	2.95×10^3	$*$
^{22}Na	6.54×10^3	$*$
^{106}Rh	655.58	
^{124}Sb	1.34×10^6	
^{44}Sc	42.56	
^{42}K	241.36	
^{88}Y	1.67×10^5	$*$

© Springer Nature Switzerland AG 2019
J. Dunger, *Event Classification in Liquid Scintillator Using PMT Hit Patterns*,
Springer Theses, https://doi.org/10.1007/978-3-030-31616-7

Table A.2 Expected uranium chain radioactivity levels in the SNO+ tellurium phase scintillator cocktail [2]

Isotope	Decay	c_{TeA} g/g	c_{LAB} g/g	c_{BD} g/g	Rate/yr	
^{238}U	α	1×10^{-13}	1.6×10^{-17}	3.5×10^{-14}	390314	
^{234}Th	β	1×10^{-13}	1.6×10^{-17}	3.5×10^{-14}	390314	
^{234}Pa(m)	β	1×10^{-13}	1.6×10^{-17}	3.5×10^{-14}	390314	*
^{234}U	α	1×10^{-13}	1.6×10^{-17}	3.5×10^{-14}	390314	
^{230}Th	α	1×10^{-13}	1.6×10^{-17}	3.5×10^{-14}	390314	
^{226}Ra	α	1×10^{-13}	1.6×10^{-17}	3.5×10^{-14}	390314	
^{222}Rn	α	1×10^{-13}	1.6×10^{-17}	3.5×10^{-14}	390314	
^{218}Po	α	1×10^{-13}	1.6×10^{-17}	3.5×10^{-14}	390314	
^{214}Pb	β	1×10^{-13}	1.6×10^{-17}	3.5×10^{-14}	390314	
^{214}Bi	β	1×10^{-13}	1.6×10^{-17}	3.5×10^{-14}	390232	
	α				82	*
^{214}Po	α	1×10^{-13}	1.6×10^{-17}	3.5×10^{-14}	390232	*
^{210}Tl	β	1×10^{-13}	1.6×10^{-17}	3.5×10^{-14}	82	*
^{210}Pb	β	1×10^{-13}	6.11×10^{-25}	3.5×10^{-14}	427368	
^{210}Bi	β	1×10^{-13}	3.78×10^{-28}	3.5×10^{-14}	427368	
^{210}Po	α	1×10^{-13}	4.15×10^{-24}	3.5×10^{-14}	1.74×10^7	

Table A.3 Expected thorium chain radioactivity levels in the SNO+ tellurium phase scintillator cocktail [2]

Isotope	Decay	c_{TeA} g/g	c_{LAB} g/g	c_{BD} g/g	Rate/yr	
^{232}Th	α	5×10^{-14}	6.8×10^{-18}	3.5×10^{-15}	49190	
^{228}Ra	β	5×10^{-14}	6.8×10^{-18}	3.5×10^{-15}	49190	
^{228}Ac	β	5×10^{-14}	6.8×10^{-18}	3.5×10^{-15}	49190	*
^{228}Th	α	5×10^{-14}	6.8×10^{-18}	3.5×10^{-15}	49190	
^{224}Ra	α	5×10^{-14}	6.8×10^{-18}	3.5×10^{-15}	49190	
^{220}Rn	α	5×10^{-14}	6.8×10^{-18}	3.5×10^{-15}	49190	
^{216}Po	α	5×10^{-14}	6.8×10^{-18}	3.5×10^{-15}	49190	
^{212}Pb	α	5×10^{-14}	6.8×10^{-18}	3.5×10^{-15}	49190	
^{212}Bi	β	5×10^{-14}	6.8×10^{-18}	3.5×10^{-15}	31482	
	α				17708	*
^{212}Po	α	5×10^{-14}	6.8×10^{-18}	3.5×10^{-15}	31482	*
^{208}Tl	β	5×10^{-14}	6.8×10^{-18}	3.5×10^{-15}	17780	*

Table A.4 Expected external radioactivity levels in the SNO+ tellurium phase [2]. The rate for the external UPW includes events up to 8.5 m from the detector centre

Origin	Contamination g/g		Isotope	Rate/yr	
Internal calibration ropes			^{214}Bi	4966	
	2.8×10^{-10}	g^{238}U/g	^{210}Tl	1.04	
	2.0×10^{-10}	g^{232}Th/g	^{208}Tl	418	
	637.06×10^{-10}	gK/g	^{212}BiPo	743	
			^{40}K	2.8×10^4	
Hold down ropes			^{214}Bi	4.06×10^6	*
	4.7×10^{-11}	g^{238}U/g	^{210}Tl	853	
	2.27×10^{-10}	g^{232}Th/g	^{208}Tl	2.32×10^6	*
	871.9×10^{-9}	gK/g	^{212}BiPo	4.13×10^7	
			^{40}K	1.89×10^8	
Hold up ropes			^{214}Bi	8.34×10^5	*
	4.7×10^{-11}	g^{238}U/g	^{210}Tl	175	
	2.27×10^{-10}	g^{232}Th/g	^{208}Tl	4.78×10^5	*
	871.9×10^{-9}	gK/g	^{212}BiPo	8.5×10^5	*
			^{40}K	3.9×10^7	
AV inner dust			^{214}Bi	4.15×10^4	*
	1.0×10^{-12}	g^{238}U/g	^{210}Tl	8.7	
	1.0×10^{-12}	g^{232}Th/g	^{208}Tl	2.48×10^4	*
	7.32×10^7	gK/g	^{212}BiPo	4.41×10^4	*
			^{40}K	9.4×10^5	
AV			$^{?14}$Bi	1.28×10^7	*
	1.1×10^{-6}	$g^{?38}$U/g	^{210}Tl	2682	
	5.6×10^{-6}	g^{232}Th/g	^{208}Tl	1.50×10^6	*
	0.01	gK/g	^{212}BiPo	2.67×10^6	*
			^{40}K	7.32×10^7	
AV outer dust			^{214}Bi	7.75×10^6	*
	1.0×10^{-12}	g^{238}U/g	^{210}Tl	163	
	1.0×10^{-12}	g^{232}Th/g	^{208}Tl	4.6×10^5	*
	7.32×10^7	gK/g	^{212}BiPo	8.2×10^5	*
			^{40}K	1.76×10^7	
External UPW			^{214}Bi	1.32×10^8	*
	2.06×10^{-13}	g^{238}U/g	^{210}Tl	2.77×10^4	
	5.2×10^{-14}	g^{232}Th/g	^{208}Tl	3.92×10^6	*
			^{212}BiPo	6.96×10^6	
PMT			^{214}Bi	3.7×10^{11}	
	100×10^{-6}	g^{238}U/g	^{210}Tl	7.87×10^7	
	100×10^{-6}	g^{232}Th/g	^{208}Tl	4.4×10^{10}	*
			^{212}BiPo	7.8×10^{10}	

Table A.5 Expected $\alpha - n$ backgrounds in SNO+ tellurium phase [2]

Name	α Emitting material	Absorbing isotope	Absorbing material	
Telab_13c	Scintillator	^{13}C	Scintillator	*
Alphan_Telab_Avin_Av_13c	AV inner surface	^{13}C	AV bulk	*
Alphan_Telab_Avin_Ls_13c	AV inner surface	^{13}C	Scint cocktail	*
Alphan_Telab_Avout_Av_13c	AV outer surface	^{13}C	AV bulk	*
Alphan_Telab_Avin_Av_18o	AV inner surface	^{18}O	AV bulk	*
Alphan_Telab_Avout_Av_18o	AV inner surface	^{18}O	AV bulk	*

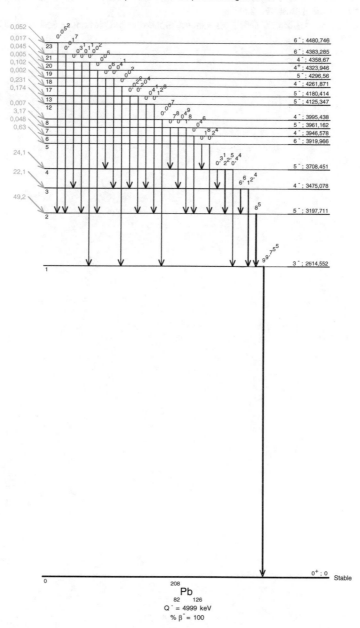

Fig. A.1 ^{208}Tl decay scheme

References

1. Kaptanoglu T (2016) Te diol 0.5% loading approved bb sensitivity plots and documentation, SNO+-docDB 3689-v2
2. O'Keeffe H, Lozza V (2017) Expected radioactive backgrounds in SNO+, SNO+-docDB 507-v35

Appendix B
The Solid Angle Correction for Vertex Reconstruction

This appendix discusses the subtleties of including explicit solid angle and attenuation corrections in the reconstruction likelihood for the slow-scintillator detector outlined in Sect. 9.2.

The likelihood presented there included a solid angle correction of the form:

$$\frac{\Omega^i}{4\pi} \tag{B.1}$$

where Ω^i was the solid angle of the front face of PMT i, viewed from the vertex position. Later, PDFs were built (e.g. Fig. 9.11) for the likelihood using the time residuals and $\cos\theta^i$ of hits produced in MC events. Hits with smaller solid angles are less likely to make it into that PDF and so the shape of the PDF already contains information about solid angle effects. The reason these factors require further thought is that it seems as though the correction has been double counted: once through the PDF shape and once through the explicit correction.

To remedy this, one could weight each entry in the PDF, P, by the inverse of its solid angle. This ensures that the PDF will converge on the undistorted PDF, P_{em}, in the limit of a large number of events.

However, it is simple to show that solid angle corrections have negligible effect on the shape of the PDFs, i.e. P_{em} and P are the same, provided the PDF is built from MC events where direction of the events is independent of their positions.

To see this, suppose the PDF is generated from N_{events} labelled $\{i = 0, 1, 2, \ldots, N_{events}\}$, each with hits labelled $\{j = 0, 1, 2, 3, \ldots, N_{hits}^i\}$. If the solid angle of hit j in event i is Ω^{ij} then the expected content of any given PDF bin, α, is:

$$C^\alpha = \frac{\sum_{i=0}^{N_{events}} \sum_{j=0}^{j=N_{hits}^i} P_{em}^\alpha \frac{\Omega^{ij}}{4\pi}}{\sum_{i=0}^{N_{events}} N_{hits}^i} \tag{B.2}$$

which is just P_{em}^α adjusted by a scale factor S^α

© Springer Nature Switzerland AG 2019
J. Dunger, *Event Classification in Liquid Scintillator Using PMT Hit Patterns*,
Springer Theses, https://doi.org/10.1007/978-3-030-31616-7

Fig. B.1 Effect of the solid
angle correction on 2.5 MeV
electron direction fits

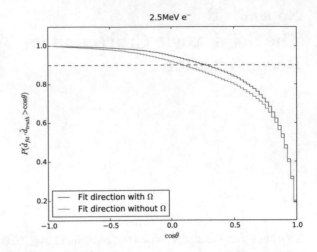

$$S^\alpha = \frac{1}{4\pi} \frac{\Sigma_{i=0}^{N_{events}} \Sigma_{j=0}^{j=N_{hits}^i} \Omega^{ij}}{\Sigma_{i=0}^{N_{events}} N_{hits}^i} \tag{B.3}$$

If the directions of the generated particles are not correlated with event position,
photon emission time or photon direction, S^α is, on average, the same for all bins so
it creates no shape distortion. This justifies the explicit correction term in Eq. 9.6.

Indeed, switching off the explicit solid angle correction notably widens the reconstructed direction distribution; this is shown in Fig. B.1. Exactly the same argument
can be made for the attenuation length term.

Appendix C
Tuning the Electron Reconstruction Algorithm

This appendix describes the method used to select the optimal effective velocity and absorption length for the slow-scintillator detector electron reconstruction algorithm.

C.1 Effective Speed

The effective speed used to calculate the time residuals was tuned for optimal fit performance on 2.5 MeV electrons. Figure C.1 shows that an effective velocity of 197 mm/ns gives radially unbiased fits, minimal position resolution and a maximum probability of reconstructing a direction in the forward hemisphere defined by the true direction.

Note that this is considerably faster the ScintFitter value for LABPPO of ≈ 185 mm ns^{-1}. To better understand this, Fig. C.2 shows an analytical calculation of the group velocity of LAB. The light velocity is a weakly increasing function of wavelength. This explains the unusually fast effective speed selected for LABSS, because the slow-scintillator emission peak is at 470 nm, compared with 360 nm for PPO. 197 mm/ns is equivalent to all of the light having wavelength 560 nm.

C.2 Attenuation Length

The total attenuation length for LABSS in the slow-scintillator emission range is 20–100 m (Fig. 9.3) so it is not clear what absorption length should be selected for the likelihood in Eq. 9.8.

In direction reconstruction, attenuation is most important when the Cherenkov photons have to take long paths to cause a hit, i.e. when the direction of a primary electron $\hat{d}_e$ is anti-parallel to the vertex position $\hat{x}_v$:

$$\hat{d}_e \cdot \hat{x}_v < -1 \tag{C.1}$$

© Springer Nature Switzerland AG 2019
J. Dunger, *Event Classification in Liquid Scintillator Using PMT Hit Patterns*,
Springer Theses, https://doi.org/10.1007/978-3-030-31616-7

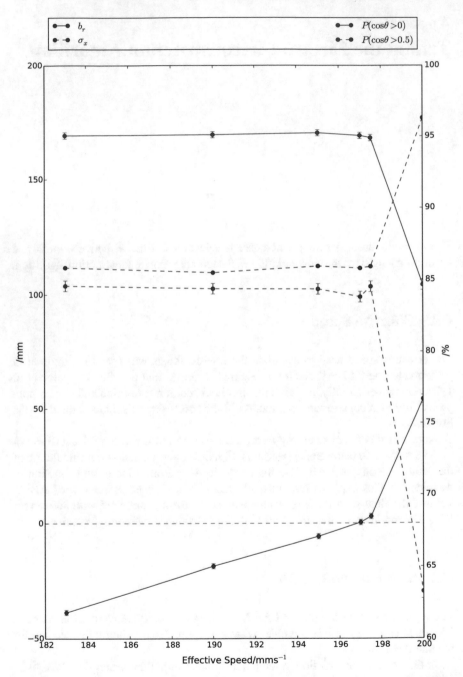

Fig. C.1 Radial bias (b_r), x position resolution (σ_x), and the probability of reconstructing above values of $\cos\theta$ against effective velocity. θ is the angle between the fit direction and the true direction

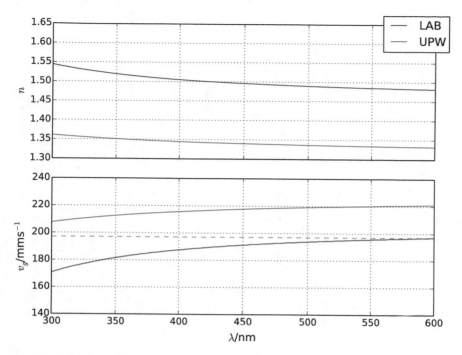

Fig. C.2 Refractive indices and analytically calculated group velocities v_g for optical photons in water and LAB. Calculated according to $v_g/c = (n - \lambda \frac{dn}{d\lambda})^{-1}$

Fig. C.3 2.5 MeV electron fit performance versus effective attenuation length

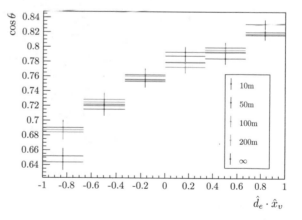

Figure C.3 shows the average $\cos \theta = \hat{d}_{fit} \cdot \hat{d}_{true}$ for several bins of $\hat{d}_e \cdot \hat{x}_v$ at a range of attenuation lengths. It is clear that an attenuation length of 50–200 m improves the reconstruction accuracy for events with large negative $\hat{d}_e \cdot \hat{x}_v$. An attenuation length of 50 m was selected for the algorithm.

About the Author

Jack Dunger studied natural sciences at the University of Cambridge before completing his DPhil at the Univeristy of Oxford, supervised by Steven Biller. His current work interests focus on new scintillation tecnologies and the application of machine learning to medical imaging and the natural sciences. He is a pianist and podcast lover.

© Springer Nature Switzerland AG 2019
J. Dunger, *Event Classification in Liquid Scintillator Using PMT Hit Patterns*,
Springer Theses, https://doi.org/10.1007/978-3-030-31616-7

Printed in the United States
By Bookmasters